What Very Fabulous People Are Saying About CODE RED

"Lisa Lister is the woman we all wish we were lucky enough to have as our sister. But short of having her on speed dial, this book is the next best thing to a full-on feminine support system!"

—KATHLEEN McGOWAN, ACTIVIST AND INTERNATIONAL BESTSELLING AUTHOR OF *THE EXPECTED ONE* AND *THE BOOK OF LOVE*

"Code Red *is a modern priestess guide for women to go within and use their feminine cycles to lead the good life. Lisa Lister sounds the call for you to heal your cycle, accept your ever-changing body, and honor your feminine soul. This is the essential foundation and practice for you to get in your 'FLO' and become the powerful force of nature that you are designed to be!"*

—ALISA VITTI, HORMONAL HEALTH EXPERT AND AUTHOR OF *WOMANCODE*

"Lisa Lister doesn't mess around. She's here to break boundaries and barriers that have been created by what society defines 'normal' and, in turn, remind us what's real. Teaching and leading other girls to honor their divinity inside and out is no easy task, and for that reason alone, I respect Lisa wholeheartedly. As far as I'm concerned, Lisa is part-angel, part-goddess, and part-rocker—she writes in a way that's friendly, honest, and captivating. This book is filled with 'aha!' moments from beginning to end. Let Code Red *help you recognize your monthly flow—as a reminder that you are fiercely feminine."*

—KYLE GRAY, BESTSELLING AUTHOR OF *ANGEL PRAYERS*

"Lisa's sheer passion is palpable, whether she is presenting her practical guide to getting the most out of your cycle or tackling complex philosophical questions of femaleness. She stands strong in that space between cycle-worship and cycle-commodification, making room for more women to feel self-knowledge through body acceptance is here for them, too. To paraphrase Gail Dines, if women loved their bodies, how much might society change for the better? Lisa gives us a glimpse of a bright future."

—HOLLY GRIGG-SPALL, AUTHOR OF *SWEETENING THE PILL: OR HOW WE GOT HOOKED ON HORMONAL BIRTH CONTROL*

"Lisa is so full of womanly wisdom; she completely blows my mind. As a very left-brain entrepreneur, I sometimes forget I have a body, let alone a female one! She reminds me to look within at the wild, messy, imperfect but oh-so-freaking powerful feminine self and use that to create a more creative and soul-fulfilling life. I can't wait to dive into Code Red *whenever I need a reminder."*

—DENISE DUFFIELD-THOMAS, AUTHOR OF *LUCKY BITCH* AND *GET RICH, LUCKY BITCH!*, AND FOUNDER OF LUCKYBITCH.COM

"Allow me to explain why I believe you should read this book. We live in a fundamentally gyno-phobic world. A world where the fear of women, (and as Freud would confirm) specifically the menstrual process with all its ramifications, has been so systemically profound for thousands of years, and where until very recently, men fully subjugated women to their will, inferring or overtly stating menstruation (as symbolic of the very nub of womanhood as it were) was disgusting, and as we know, sadly in many parts of the world, this level of supreme ignorance, fear, and oppression continue to this day. So deep runs the conditioning, women themselves have been inculcated with self-disgust on account of it. So when someone as bold, brazen, and erudite as

Lisa Lister emerges to redress the balance by turning the issue on its head and proposing women feel pride in all their aspects and functions, we must afford her all the support we can muster.

For if we are to evolve as a species now, it will be the women who lead us in it, ideally with the support and help of the men, but certainly no longer in deference to them. So all plaudits to Lisa and may this book reach all the people it needs to, to trigger the next phase of the requisite shift. And (talking as an honorary sister), may we now prevail (before it's too late)."

—BAREFOOT DOCTOR, AUTHOR OF *THE BAREFOOT DOCTOR'S HANDBOOK FOR THE URBAN WARRIOR: SPIRITUAL SURVIVAL GUIDE*

"Lisa's writing is full of solid truths, the sort that will have you nodding as you read. She speaks from her heart, her experiences, and her soul and is one of the most authentic people I know. Her knowledge, gained from living it, feeling it, and being it, asks questions and then answers them beautifully, all done in her signature straight-talking, no-nonsense style. Love, love, love."

—DAVID WELLS, ASTROLOGIST, PSYCHIC, AND AUTHOR OF *YOUR ASTROLOGICAL MOON SIGN* AND *PAST, PRESENT AND FUTURE*

"Lisa Lister is love on fire. A true embodiment of the goddess on Earth who walks the fire path of love, truth, fleshly feelings of wild desire—women wake to unconditional and healing love in Lisa's wake. Her words, all from the deepest caverns of her heart, touch the hearts of all who read her. To read, Lisa is to know her and to know Lisa is to know oneself. Her journey is yours, and yours hers, and all together, we piece the goddess back together on the planet."

—SARAH DURHAM WILSON, FOUNDER OF DOITGIRL.COM

"Every woman NEEDS to read this book. It will make you laugh, cry, and nod in agreement, but it will also give you 'aha!' moments of total inspiration. This book tells YOUR story. Lisa Lister is handing us the key to our own freedom, enabling us to find ourselves and reconnect with our own inner power, releasing us from the bonds that society, media, and even our own minds put in place. This book brings a new era... vive la menstruation revolution!"

**—RACHEL PATTERSON, AUTHOR OF *MOON MAGIC*
AND *THE GRIMOIRE OF A KITCHEN WITCH***

"An important book for our current cultural times. As women continue to fight for equality, Lisa reconnects the modern woman with her innate feminine wiles. Yes, it gets bloody and intimate, but the lack of that truth has sanitized our existence... until now. In this intriguing and beautifully unsanitary account, Lisa allows us to own our bodies and to know what and how they really work. Delving beyond the physical, this is a spiritual, transformative read. One for all our sisters and daughters."

**—ALICE GRIST, AUTHOR OF *THE HIGH-HEELED GUIDE TO ENLIGHTENMENT*
AND *DEAR POPPYSEED: A SOULFUL MOMMA'S PREGNANCY JOURNAL***

"A priceless book for girl—and womankind. This book beautifully relates the menstrual cycle phases with nature's seasons and the Moon phases. This has become my go-to manual for life planning, e.g. when to make business decisions, approach relationship questions, or take time off for resting. Jam-packed with bloody glorious information!"

**—ANI RICHARDSON, AUTHOR OF *LOVE OR DIET: LOVE YOURSELF*
AND *RELEASE THE NEED TO BE COMFORTED BY FOOD***

CODE

Red

Know your flow, unlock your

monthly superpowers,

and create a bloody

amazing life.

Period.

Lisa Lister

HAY HOUSE

Carlsbad, California • New York City
London • Sydney • New Delhi

Published in the United Kingdom by:
Hay House UK Ltd, The Sixth Floor, Watson House,
54 Baker Street, London W1U 7BU
Tel: +44 (0)20 3927 7290; Fax: +44 (0)20 3927 7291
www.hayhouse.co.uk

Published in the United States of America by:
Hay House Inc., PO Box 5100, Carlsbad, CA 92018-5100
Tel: (1) 760 431 7695 or (800) 654 5126
Fax: (1) 760 431 6948 or (800) 650 5115; www.hayhouse.com

Published in Australia by:
Hay House Australia Pty Ltd, 18/36 Ralph St, Alexandria NSW 2015
Tel: (61) 2 9669 4299; Fax: (61) 2 9669 4144; www.hayhouse.com.au

Published in India by:
Hay House Publishers India, Muskaan Complex,
Plot No.3, B-2, Vasant Kunj, New Delhi 110 070
Tel: (91) 11 4176 1620; Fax: (91) 11 4176 1630; www.hayhouse.co.in

Text © Lisa Lister, 2015, 2020

This book was previously published in 2015 by SHE Press UK and has
been substantially revised and updated for this edition.

The moral rights of the author have been asserted.

A catalogue record for this book is available from the British Library.

Tradepaper ISBN: 978-1-78817-475-6
E-book ISBN: 978-1-78817-479-4
Audiobook ISBN: 978-1-78817-484-8

Printed in the UK by TJ International Ltd, Padstow, Cornwall.

MIX
Paper from
responsible sources
FSC   FSC® C013056
www.fsc.org

Contents

Foreword

White has never been my color. It has always made me anxious. I don't like the pressure it creates. It expects too much of me. I have to play more standoffishly with my young son, eat more mindfully, and wear the right hued lingerie. So, when a close friend invited me to see her spiritual teacher, I had to scramble to find some white to wear. My little sister let me borrow an all-white cotton summer dress—a dress she would end up never seeing again.

It was the last day of my period. I didn't feel overly concerned because it was so light at that point. But the pressure of wearing white forced me to go to the ladies' room to make super-absorbent sure I was all set before getting in line to see my friend's spiritual teacher. The love in the room was sublime. As I approached him, I could feel the static-like energy of spiritual forces at work all around me. He was seated in a regal-looking chair. Once I was standing beside him, he took my hand and told me to come to see him in Brazil. I either just nodded or said, "OK," slightly dazed by the intensity of his gaze. Then I was directed to sit in a chair near him to have his "entities" work with me.

I felt so warm and peaceful. Time melted and morphed. It was unclear how long I was sitting there. I just felt this heaviness and knew I needed to stay seated. And then as soon as the heaviness lifted, I knew I could leave. So, I went to stand up, still drenched in utter bliss, and noticed an odd sensation; my dress, or rather my sister's dress, was clinging rather tightly to my backside. My first hope, or thought, was that the bucket seat of the chair and the warmth of the room must have made me sweat. But immediately after that thought, I knew what had happened. So, I sat back down.

I considered my options. I could strategically walk out of the room with my back to the wall, slightly smiling to distract attention from my odd choice of exit. I could feel shame for turning this white dress to a very conspicuous shade of red. I could make every possible effort to hide, deny, and dismiss the fact that I'm utterly female. Or, I could stand up with my head held high and walk at a calm, normal pace across this room with several hundred very spiritual people doing very spiritual things, and just let my red be a part of it.

I'm not sure I've ever made a grander exit.

Of course, I would never have contrived or created that opportunity to own my red so publicly, at least not consciously. But in the face of being ashamed or living my truth—that the body is sacred—I'll always choose to embody the latter. And I have to admit, I felt crazy triumphant when I got to the café of the spiritual center to change. I had walked out of an atmosphere so spiritual it was nearly disembodied with absolutely zero shame that I was very clearly in a body. The sheer bravado of my body still makes me laugh… the fact that the body never lies. A small piece of my power came with me when I left that room with such inviolable dignity.

It's one thing to know cognitively that I've nothing to be ashamed of for being a woman. It's something else to act based on that knowing. As a theologian, I have studied the Divine Feminine to honor and give voice to what's feminine within me and in others. I wanted to provide a sense of balance to our ideas of what's sacred. I also found validation for how I exist in the world. I move from within. I live in a deeply intuitive way, which is how the Divine Feminine works. The internal terrain is as significant and important as the external surroundings. Take, for example, the voice of the Divine Feminine in an ancient text that dates back to the 1st century titled *The Thunder, Perfect Mind*. Verse 30-31 reads, "Since what is your inside is your outside. And what you see on the outside, you see revealed on the inside."[1] So, I've made a spiritual practice of going within and trusting what that small, unassuming voice has to say. And then, I believe it enough to take action on its directives.

The Gospel of Mary Magdalene, translated from both Greek and Coptic, contains wisdom that Mary received from Jesus in a vision. She even reveals how she "sees" or receives this and other visions of Jesus. Then there's a passage at the end where Peter, aka. "the rock," doubts everything Mary recounts. His primary

---

1. The New, New Testament. Ed. Hal Taussig. "The Thunder: Perfect Mind." (Houghton Mifflin Harcourt, 2013) p.185.

peeve is that he can't believe or begin to understand why Jesus would reveal all this secret wisdom to Mary… a woman! In other words, the sheer fact that Mary Magdalene is a woman should have invalidated her from being the one to receive these visions. And yet, somehow, it made her the perfect disciple because she could see Jesus in this very inward, mystical way. Levi comes to Mary's defense and demands, "If the Savior considered her worthy, who are you to disregard her?"

I see *Code Red* and what Lisa is creating as a part of a zeitgeist, a rising, a reparation, a 21st-century response to this 1st or 2nd-century question. Who are we to disregard the divine feminine? On a soul level since I was a little girl, it has felt like a sacred obligation to honor the feminine attributes in men and women; to respect the wisdom of the human body; to acknowledge the inherent worth and dignity of being female; to remember the holy power and healing that's innate to the sexual energies we all contain. And it has felt so crucial to acknowledge the sacred in what we've considered mundane, or even shameful, about our very human lives.

The Swiss psychologist Carl Jung wrote extensively about alchemy. Alchemy is an ancient practice of transmuting base metals into gold. It's also a metaphor for the psychological process of transforming pain into love, of becoming whole. The fourth and final stage of the alchemical process is called the Rubedo, which means the redness or the reddening. It's when the ego merges with the self, when the light and dark within us unite, or when our suffering finally turns into much-needed wisdom. The Rubedo is when what's sacred and timeless can be here with us in this fleeting, everyday moment. This alchemical stage is what *Code Red* gives us the spiritual tools to be able to do—to walk with inherent worth and dignity as a powerful priestess no matter how messy or human the moment we find ourselves in.

Meggan Watterson, author of
Mary Magdalene Revealed: The First Apostle,
Her Feminist Gospel & the Christianity We Haven't Tried Yet,
REVEAL: A Sacred Manual for Getting Spiritually Naked,
and *How to Love Yourself (And Sometimes Other*
People): Spiritual Advice for Modern Relationships

Introduction

I bleed every month for five days and
don't die. Seriously, don't tell me
I'm not a fucking superhero.
–Lisa Lister (day 25 of her cycle)

So, there's a code.

It's ancient, it's deeply spiritual, and, more than that, it's powerful.

Fiercely powerful.

So powerful that it has barely been spoken for 3,000+ years.

Menstruation.

Yep, your period is way more than the PMS, carb cravings, and fierce rage that you may experience each month. Your menstrual cycle/period/bleed [*insert whatever you choose to call your "time of the month" here*]–is actually a four-part code that, once cracked, will uncover a series of monthly superpowers. You can use these to enhance your relationships with lovers and others, build a better business, have incredible sex, and create a "bloody" amazing life.

Until now, your monthly bleed might simply have indicated whether you were pregnant or not. You may get such painful cramps and suffer so much discomfort that the idea of finding love for your cycle makes you want to punch me in the face. And many of us have been taught it's shameful and/or embarrassing to bleed. Tampon adverts tell us our period is "Mother Nature's curse." Some advertising campaigns turn our blood into blue water to sanitize the process, to distance and disconnect us from our own blood. While others have made our bleedtime look like it's the perfect opportunity for high-energy pursuits like rollerblading, sailing, and riding rollercoasters while wearing an all-white ensemble.

It's no wonder so many of us have such resentment toward our menstrual cycle, which is why I totally get that beginning to appreciate or even, dare I say it, *enjoy* our menstrual cycle, is a really big ask.

But hear me out.

There was a time when women totally honored and celebrated their monthly cycles.

For real.

It was the cycle that moved a woman through the archetypes: from girl to mother to wise woman through to crone.

Back in the day, when girls began to bleed, a beautiful menarche ceremony would be held to initiate her into womanhood.

She was celebrated.

Her bleed cycle became a part of the natural rhythm of her life, and she'd gather together with the women of her family and community to bleed at the dark phase of the Moon. This was her time to go inward, to reconnect with herself, her sisters, and Mumma Earth. This quiet time, dictated by her body's rhythms, refueled her, filled her up, so she could pretty much rock out in the weeks ahead.

In my own traveler culture, men knew that women were at their most powerful during their bleedtime, so took care not to offend a woman through fear of being rendered *mokadi*—having something of his soul taken away. Basically, these men knew we were badass and capable of serious magic-making when we bled so kept their distance. Ha!

Our cycle—the bleed, and everything that went with it—was sacred.

But for over 3,000+ years now, our stories, our truths, and our wisdom as women have been distorted, censored, burned, and unheard because those in charge of creating the systems and structures got fearful. Christianity changed up its teachings to totally disregard the potent power of our menstrual cycle, (along with a fair few other things along the way). Because of this, we lost our daily connection to our body, to Mumma Earth, the Moon, the seasons, and their cycles.

Now we reside in a society with masculine constructs that has no guidance or structure for the feminine menstrual experience.

This complete disconnection from the cyclic experience, from our power source, has caused many of us to hate, cut, add to, and mutilate our bodies as we endlessly strive for an advertiser's idea of so-called perfection. We ignore our deepest needs as women because we no longer trust that we know ourselves better than anyone else. We look outside ourselves for the answers. We spend time and money on self-help books, personal-development courses, and feel-better workshops, trying to find a way, a spiritual practice that helps us to makes sense of ourselves as women. Yet we find nothing that actually fits. When we get emotional or dare to feel, we apologize for our tears, suppress our anger, and fear being called hysterical. And when that happens, we may self-medicate with drugs/food/alcohol/shopping.

Worst of all, we have an epidemic of "down there" pain and disease—PMT, polycystic ovaries (PCOS), endometriosis, fibroids—and overwhelm, stress, anxiety, and fertility issues are at an all-time high. We manage our menstrual cycle with synthetic hormones, denying ourselves the experience of living fully in our power. In essence, we are stripping Wonder Woman of her lasso.

This is a *Code Red* situation, and I want you to consider this book your call to action.

A rallying cry to any woman who has ever:

- found her cycle annoying and an inconvenience

- been disgusted by her blood and felt dirty and shame-filled at menstruation

- wanted to hide away and pretend menstruation didn't exist

- talked about her period in hushed tones

- been crippled by premenstrual or menstrual-related pain and dis-ease

- hated it because her bleed meant that another month had passed and she wasn't pregnant when she desperately wanted to be

- thought: *There must be a freaking reason as to why we bleed for five or more days each month and don't die?!*

It's no secret that most women aren't entirely happy with menstruating, but that's because most women don't know that their cycle has a much bigger purpose than they have been led to believe.

As women, we will experience 350-500 menstrual cycles in our lifetimes. *Code Red* will dare you to get to know yourself better through the wisdom, guidance, and insight that each cycle provides.

Each cycle takes us on a journey through both the light and the dark parts of ourselves—the masculine and feminine; the yin and yang. As we move through each of the four phases—pre-ovulation, ovulation, premenstruation, menstruation—we get the chance to renew and refresh our entire being—physically, mentally, psychologically, and spiritually—EVERY FREAKIN' MONTH.

That's pretty cool, right?

Now, I'm not a medical doctor, but I am a woman who hated her period. In fact, 15 years ago, I hated everything about my life.

I'd binge in secret and hide the wrappers down the side of the sofa. I strived so hard to "be" and "do" in a job I *thought* I wanted, but instead of feeling success and happiness, I became jaded and stressed. I was in a relationship that was going nowhere, yet I didn't dare to demand more. And my periods were heavy, all-consuming, and hurt like a motherlover; sex was painful, and orgasms were non-existent.

Life was pretty sucky. (Total understatement.)

After three years of misdiagnosis, I was told I had severe endometriosis and needed surgery to remove my womb. At this point, I'd never heard of endometriosis. I also had NO love for my lady parts, and the idea of having my whole down-there-ness removed was momentarily tempting. But walking out of that hospital appointment with a handful of literature as to the next steps, I felt a pain in my belly. This wasn't the pain I'd previously experienced, but fiery anger stirring in my womb, my ovaries, and my vagina.

She was pissed, and it was at that moment that I knew a hysterectomy wasn't an option. In fact, so driven by deep anger at modern medicine's quick-fix "whip it out" response to women's health, my womb, ovaries, and vagina took me on a 10-year exploration of my own lady landscape.

It's been quite an adventure, and through writing *Code Red* and reliving some of it within these pages, I've found something far more powerful and potent than modern medicine: menstrual medicine. A deeply cleansing and healing self-love practice that works predominantly with the creative, spiritual, and psychological energies of the four phases of the menstrual cycle. It has become the very essence of IN-YOUR-BODY-MENT, a movement practice that I teach and share with women both in-person and online, and has brought me into a rather incredible relationship with my lady landscape, my menstrual cycle, my body, and myself.

This shouldn't be something we have to discover; it's our birthright as a woman. The menstrual cycle, if we work with it and not against it, is an incredibly potent experience, and my big heart wish for this book is that it's your call to power.

I now live my life totally in sync with my cycle: my business, relationships, sex life, health, money, and creativity are all in complete alignment with my rhythms. And the good news? I still have my womb, despite the doctor's best efforts to take it out. My endometriosis is currently manageable—yes, it's possible and through finding fierce love for my cycle, I've found love for myself as a friend, as a lover, as a Creatrix, as a woman.

I want this for you, too.

I want this for EVERYONE that bleeds.

Code Red is my invitation to you to explore, navigate, and love your lady landscape. It's a book, it's a self-inquiry practice, and, most importantly, it's your new best friend. An opportunity to reconnect with your true nature as a woman, to use your menstrual cycle as an ever-unfolding map to crack your code and create a bloody amazing life.

Big, big love…

Lisa x

How to Use This Book

As I mentioned, I'm not a doctor, and what I share isn't intended to serve as medical advice or to diagnose and treat any disease or serious health condition. Personal responsibility is most definitely required.

In fact, everything I share is from my lived experience and journey as a woman with a menstrual cycle and a womb, ovaries, cervix, and vagina in my physical body. This may not apply to you, and that's totally OK.

For those of you in menopause, if you no longer bleed, you don't have a womb, you've had early induced menopause, or any of the many other reasons as to why you don't experience a menstrual cycle, this doesn't in any way mean that you're without power–far, far from it.

The same applies to those who take and use contraception, does this affect the cycle? Yes. Does it mean you can't explore what I share in this book? Absolutely not. I share more on this in *Chapter 1: What's Really Going on "Down There"?*

I talk of women throughout this book because it's empowering to connect to the menstrual cycle within the framework of feminine wisdom and matrilineal power. I want to acknowledge, however, that I'm aware that not all women menstruate and not all those who menstruate are women. I welcome opening up the conversation for all of us who menstruate, regardless of gender identity.

Finally, "SHE" is very much a word I use as my personal expression of the Divine, and "lady landscape" is how I refer to my vulva, vagina, womb, and ovaries. If these don't work for you, please feel free to replace them with words or phrases that do.

This book is simply an invitation for self-exploration—take what feels right, and leave the rest. Sound good? Then let's ride the crimson wave...

CHAPTER 1

What's Really Going On "Down There"?

Our menstrual cycle is powerful. It impacts the way we think and feel from one day to the next, not just during our bleedtime, but also throughout the entire 28 to 30 days. (Yours may be much longer or much shorter—our cycles are totally unique to us.)

It's also true that we have strengths—what I call SHE powers—during each phase of our cycle. Unfortunately, we haven't been taught the intricacies of these powers and how we can *really* leverage them.

Well, 'til now that is.

Some cycles map the most auspicious time for everything in life, and we're fortunate enough to have an incredible internal map, our menstrual cycle, that gives us directions on the most promising/supportive/helpful times for everything we do. Unfortunately, Western culture decided that women's inner guidance system—our very own GPS—is obsolete and reprogrammed us to follow a seven-days-a-week calendar, instead of our own bodies and the timing of nature. Yep, for thousands of years, a woman's menstrual cycle was part of the natural rhythm of life.

So what happened?

Well, for starters, Christianity changed up its teachings to totally disregard the potent awesomeness of our bleed cycles. We've become more technology-obsessed, we're busy, we're tired, we've lost our daily connection to *Shakti* (our vital creative life force), Mumma Earth, the Moon, the seasons and their cycles. We've been seduced into thinking that PMS, pain, or any other lady landscape dis-ease is "normal," something we have to endure, something that simply comes with the territory of being a woman.

Yet, we don't know the territory. There's a whole part of our lady landscape that's been left unexplored, and our experience of being a woman is totally out of whack because we're working *against* our cyclic nature and causing emotional and physical trauma in our bodies.

Basically, we're no longer letting our cyclic nature guide us. Instead, we're letting the pain and dis-ease we experience each month control us. We're determined to prove that if we can just do it all—Mummahood, career, relationships—we'll be worthy of admiration, adoration, and respect.

The thing is, you *are* worthy of admiration, adoration, and respect RIGHT NOW, simply by showing up, exactly as you are.

It's just that we now live in a solar-based, masculine-led society—one where the emphasis is to continually strive and do and be successful. Except as women, this isn't a place that's truly comfy for us, at least not all of the time.

Feminine energy is fluid, it's not consistent. This means at each phase of our menstrual cycle, we show up to life differently.

As women, we're not linear. We're not meant to go, do, and act with big, vibrant energy ALL the time. As we move through the different phases of our menstrual cycle, managed by different hormones, we move through different energy levels, moods, and needs.

Cycle power

The first half of your cycle, the follicular phase, is masculine, and it's typically a time for big energy, amped up creativity, and when you're much more able to talk a good talk and to think in straight lines. The second half of your cycle, the luteal phase, is feminine, a time when you begin to move inward and want to withdraw from the world. You may crave quiet and feel a deep need for contemplation.

Unfortunately, very few of us even know this about ourselves. If we do, we probably believe we'll be less productive, less useful if we allow ourselves to follow the rhythm of our cycles, so instead, we try our best to "do it like a dude." We try to maintain the masculine, go-for-it, I-can-do-it energy throughout our whole cycle, but by doing so, we're ignoring our fundamental needs as women. Right now, many of us are barely surviving the second half of our cycle.

This is why through the pages of *Code Red*, I am going to dare you to get intimate with your menstrual cycle. I want you to claim your *entire* cycle—its intricacies, its superpowers, its teachings— and to use it as your SHE power source.

I'm so passionate about sharing this menstrual mistress-ry with you because I don't want to be the only one living a bloody amazing life. I was told I'd have to have my ovaries and womb whipped out to "cure" my endometriosis pain at the age of 25. Yet by knowing my flow—through unlocking my monthly superpowers and working with my cycle—my womb and ovaries are still in working order over 15 years later. Leaving me, to use the superpowers of my cycle to navigate and co-create a bloody amazing life.

There's so much more juice to be squeezed from life when you live in sync with your cycle, which is why I'm so passionate for you to discover this for yourself. If you thought charting your cycle was only for women who want to get pregnant, think again. I share a downloadable and not-at-all-boring-I-promise charting system to help you to identify, recognize, and use the most appropriate superpowers available to you at any given moment; a far more efficient use of time and energy than trying to be a do-it-all superwoman. You'll instead become SHE-rarr (costume is optional; I don't know about you, but I LOVE dressing up), a superhero led by the powers of her womb. Oh yeah.

As with life, my approach to this work is to keep it simple, keep it sacred, and keep it **SASSY: Spiritual, Authentic, Sensual, Sensational YOU**.

SASSY is my personal formula for becoming the mistress of your destiny. Like the chakras, when all elements of your SASSY are aligned, the awesomeness of all the women that have gone before and all the women that are yet to come—priestesses, sorceresses, seers, badasses, and wise women—is unlocked within you, and gives you an access-all-areas pass to your SHE Power. When applied to your cycle, we can use the elements of SASSY to understand the superpowers that each phase holds and how they manifest within you.

I've also tried to keep the science talk to a minimum. Not because it isn't important, it is—it really is—which is why I have been sure to share insight into how our hormones work in each phase of the cycle. Still, for me, *Code Red* is calling us back to our bodies, to reconnect with our ability to feel and to trust, which is why I am keen to encourage you to get out of your head and into your womb space and *feel* into how you show up throughout your cycle. Although if you do get off on all things science-y, and there's so much deeply important work that's being done in the hormonal and menstrual health field, do check out the REDsources section at the back of this book for more details. The stories of women are at the center of all the work I do, which is why I've invited some of my most favorite women—clients, peers, mentors, and teachers—to share theirs with you throughout the book.

Why? Because we should all be talking about our cycles and our superpowers over wine with our girlfriends—and the men in our lives should know about it too. My hot Viking husband LOVES knowing about my cycle. I've been geeking out about it for so long now, I come with my very own manual—which is obviously subject to change because that's the beauty of being a cyclic, non-consistent, ever-changing woman. But ultimately, he now knows there are specific days and times in each month that he can either avoid or work to his advantage. For example, days 7 and 8 of my cycle, I'm positively vixen-like and orgasm-inducing; they are drop-everything-and-take-me-to-bed days. But on days 24 and 25 of my cycle, I'm Kali Ma incarnate and it's best he throws chocolate at me from a distance and works the late shift.

I've seen so many men throw their arms to the sky and ask, "Why don't women come with manuals?" Well, we do: it's our menstrual cycle. And once you begin to honor and respect your cyclic nature by getting to know it intimately and sharing your findings with the men in your life (dad/husband/brother/son), not only do they get a better understanding of you but also how to honor and respect the other women in their lives too.

You may feel resistance to what I have to share. In fact, I can pretty much guarantee you will feel at least *some* resistance.

I've had to become really OK with clients not liking me a lot when we first start working together. Particularly in SHE sessions, where we talk intimately about the issues of menstrual health, fertility, and our relationship with our down-there-ness.

You may want to numb it, push it aside, or pretend it doesn't exist.

You may be physically repulsed by some of my suggestions.

You may be pissed at me because you just want a fix-it solution as to how you can enhance your fertility and get pregnant. (Working with the information in this book *can* enhance your fertility, but it's about *so* much more than that.)

You may think that it doesn't apply to you because you have endometriosis/polycystic ovaries/fibroids, or because you're perimenopausal or have gone through the menopause.

Well, it DOES apply to you.

We owe it to womankind to know ourselves this intimately, to know about the potential and the power held in the menstrual cycle and then to share what we find out about ourselves with everyone we know. Because it's in the sharing of our stories that our truth is revealed.

And, just so you know, everything I'm going to share with you, you know already.

Except right now, you've mislaid the key.

This isn't a new concept or another modern health trend, this is ancient wisdom passed down through our blood, and every time we bleed, we have access to its power. Basically, this shit is sacred.

Sacred doesn't mean it's inaccessible though.

My job, as menstrual maven, is to make this wisdom totally accessible and, most importantly, relevant to you as a modern woman.

Your menstrual cycle is your unique-to-you code that, once cracked, will give you access to *everything* you need to create a bloody amazing life.

This is why, in answer to the awesome woman that emailed me saying: "Lisa, I don't like all this talk of blood and periods. I liked it when you wrote about women's bodies, self-esteem, and empowerment, will you go back to that at some point?"

I AM writing about that.

Talking about, exploring, and getting to know our cyclic nature, our bleed cycle, and our body parts are EVERYTHING to do with our bodies, self-esteem, and empowerment.

It's the ultimate feminine power.

It's SHE power.

It's the kind of power that's going to change the world.

And FYI: If you get to know yourself through your cycle, you'll never need to read another self-help book, EVER.

Just saying.

How well do you know *your* cycle?

If I were to ask you which day of your cycle you're on, would you know? More importantly, would you even care?

I really began to get intimate with my cycle in 2004 when, after three years of totally debilitating pain and misdiagnosis, I was told I had endometriosis. If I'm honest, I was relieved. I finally had a name for the thing that had made me pretty much unbearable to live with or be around. Except, when we name things, we give them power, and the power of endometriosis among the medical profession at the time was that it gave an insta-green light to suggest taking out my ovaries and womb to "cure" the pain.

I was so shocked.

I was 25. My boyfriend and I hadn't even had the "children" chat yet.

Their response? "Endometriosis means babies are no longer an option, so why would you need your womb and ovaries? Surely being pain-free is the goal?"

Sure, the pain was debilitating; I had to take myself to bed for entire days, and I was forever canceling meetings, appointments, and social engagements.

In fact, I got myself a reputation as a total flake just because I was too embarrassed to say, "I can't come out/to work/to the meeting because I'm bleeding through industrial super-size pads and my bed looks like a scene from the movie *Carrie*."

But—and I didn't know the incredible power that we hold in our womb space at the time—SHE awoke in me, a fierce rage that meant I simply wasn't going to give it up.

I did a LOT of research. I read books and read about other women with endometriosis. Each case was different and, most importantly, I discovered that having endometriosis didn't necessarily mean I could no longer have children.

I got mad at modern medicine's quick-fix "whip it out" mentality, and I also got pretty mad at the boy, masquerading as a man, who I thought was to be my forever love. For him, the endometriosis was just one big, "bloody" inconvenience.

Not only could I not have sex as often as he liked, when I did, it was so painful that I just wanted it over with. It wasn't fun, there were very few orgasms, and there was a sense of ridiculous obligation on my part and increasing disappointment on his.

As you can imagine, this did not a long-term relationship make. He was most definitely *not* the forever love, and after reading *Eat, Pray, Love* by Elizabeth Gilbert in one sitting, we were over. It's official: books change lives.

And then I went on an adventure.

Exploring my lady landscape and what it meant to be a woman.

Shit got deep.

I took part in magic rituals, tried vaginal steams for the first time (way before Gwyneth Paltrow made it cool/controversial), stood in a circle with incredible women, had healings from shamans and womb women, danced ecstatically, initiated myself to SHE by immersing naked in the White Spring in Glastonbury, drank sacred cacao, was introduced to womb and abdominal massage, experienced and loved womb yoga, ate differently, took all kinds of incredible and interesting herbs and tinctures... and slowly began to realize that I had my very own inner GPS system—my womb. I realized that if I dared to trust it, I could begin to heal my own pain. While modern medicine is great, and it saves lives, modern attitudes surrounding it, however, are not.

I let my womb lead me to different practices and healing modalities, and to the books I needed to read.

Which is how, after reading *The Pill: Are You Sure It's for You?* by Jane Bennett and Alexandra Pope, I had a MASSIVE revelation about how I was totally dishonoring my power as a woman by medicating my bleed.

Now, I'm no Judge Judy, so there's absolutely no judgment at all if you choose to take the pill or any other hormone-based form of contraception. But if you *do*, I urge you to at least hear me out.

I was no stranger to replacing my hormones with synthetics myself. Before I had the contraceptive implant, I was on the pill.

So for 15 of my menstruating years, I'd been medicating my bleed with a pill that makes levels of globulin, the stuff that binds testosterone and affects our libido, four times lower, forever.

If that wasn't scary enough, I had no connection AT ALL with my cycle. Anything that pumps synthetic hormones into your body—the coil, implant, the pill—*will* affect your natural rhythms. As well as stopping babies from being made, it creates a synthetic balance that numbs the ebb and flow of your cycle. Basically, if you're currently on the pill—that thing that you've been calling your period? It isn't one at all.

And what I know now that I didn't know then, is that when you're not able to connect with your cycle, you're not connecting with your true nature.

Taking the pill watered down my entire experience of being a woman: the cyclic ebb and flow of emotions, access to my superpowers, and the energies that are you-nique to each of us and become heightened at specific times of the month. I became totally senseless to the wisdom my body was trying to share with me just so I didn't feel pain or inconvenient emotions, so I could mask my bleeding from both myself and others and function more "normally" in the world.

For normal, read "masculine."

Seriously, it's no wonder endometriosis, fertility issues, low libido, PMS, fibroids, and mood swings are so prevalent. In essence, we've silenced our direct hookup to our cyclic nature, to our ultimate SHE power.

Many of us know *nothing* about our cycle.

And by not knowing about it, we're not talking about it.

And by not talking about it, we're perpetuating the myth that our bleed is "wrong" or "shameful" and that a huge-ass part of being a woman isn't cool.

We deny ourselves the opportunity to truly know, protect, claim, embody, and, most importantly, love being a woman.

That chainsaw-ripping-at-your-insides PMS pain? The swollen, achy breasts? They're for a reason.

Your reaction to the coil? It's a message from your womb.

So, the more cycle-savvy you get, the easier it will be to hear your body's wisdom and live in total alignment with your bleed cycle.

Coming off the pill, by Jane Bradley

Since I started menstruating, I must have had over 200 periods. That's a lot of blood, a lot of pain, and a lot of hours gnashing my teeth, cursing the "curse" and attempting to console myself with chocolate, hot water bottles, and the Harry Potter boxset, all while double-dropping painkillers to a soundtrack of self-pitying sobs.

For a long time my periods were monthly torture. There was definitely no rollerblading while wearing white jeans. Instead, I was clumsy, anxious, achy, and uncomfortable, and so drained I used to have monthly melodramatic fantasies about collapsing while walking home from school because I was too tired to take even one more step. Like a clichéd advert for tampons or ice cream, I'd be an inexplicable ball of fury one second and a crying mascara mess the next. And that's before the bleeding or pain started.

I had a heavy, irregular flow from the start, so I lived in fear of leaks. That terror, coupled with the cramps, backache, and upset stomach that accompanied every bleed, meant I ended up staying off school for a day or two each month. Once the pattern had been recognized, it was off to the doctors.

That's how I found my teenage self taking Microgynon 30, an oral contraceptive pill that classes depression, weight gain, and headaches among its common side effects. Prescribed to regulate my cycle and manage period pain, I ended up taking it on and off for years.

On my artificially regulated cycle, my flow was lighter than it had been but still heavy. The pain wasn't as extreme but still bad. To start with my mood swings seemed to "even out" to some extent. The situation was still far from ideal, but overall it was an improvement, and it's not nicknamed the "curse" for nothing, right?

And then, very gradually—so gradually it took me years to identify—the initial improvements started to diminish. And once they disappeared altogether, they were replaced by other symptoms and side effects. The pain was as bad as it had ever been. My weight crept up. I felt numb to positive emotion—it was like I was in a bubble. While previously I'd been overwhelmed by everything, emotional and on edge for the days before and during my period, that now seemed to be the case almost permanently, intensifying to severe depression while I was actually bleeding.

During this slow process I'd been to the doctors, but their only answer was antidepressants, which I didn't want. I still wasn't sure if the pill was even responsible for how I felt, but I knew the situation was spiraling and that something had to give.

For me, the penny dropped when I read a throwaway comment in one of Elizabeth Wurtzel's books, where she describes coming off the pill as "coming up for air." I decided to do the same, and after a turbulent few months while my body adjusted, that was exactly how I felt. Although I'd had my suspicions that the pill might be making me ill, it wasn't until I'd actually ditched it that I realized how dead inside I'd felt while taking it.

Once I came off the pill, though, I had a problem: after years of chemical interference, I had no idea what to expect once I started letting my body do its own thing.

I did some research into the various stages of the menstrual cycle, and the impact it can have on creativity, energy levels, sex drive, and everything in between. I downloaded a simple menstrual calendar app and began keeping track of my bleeding, plus any other physical or emotional effects I observed throughout the month. Before long, I had an overview of what stage I'd be at and when.

Other than the obvious convenience of not being caught out by bleeding, being able to see on my calendar when the monthly emotional maelstrom will probably be at its worst has been an absolute revelation.

Turns out being able to differentiate between menstrual mood swings and an impending mental breakdown is really, really useful. Who knew?

Now that I can identify and interpret the emotional and physical processes of my cycle, I can work with, rather than against, my body. I schedule big projects and meetings for the times of the month that I feel at my most creative and invincible, rather than when I'm more likely to need reflection and rest. It's a process and I'm always learning, but so far it has made a massive difference.

The more I've got to know my flow, the more positive changes I've been able to make. I've mostly ditched dairy now, and it's made my periods much, much lighter, meaning I worry a lot less about leaks and feel less wiped out. And while feeling fragile and exhausted used have me reaching for comfort foods, I've found that steering clear of sugar, caffeine, and alcohol while I'm on my period has smoothed out my energy levels so they don't spike then crash, leaving me feeling worse.

It's taken me a long, long time to get to this stage (it's more than three years now since I came off the pill) and there have been some steep learning curves. But by getting to understand my cycle rather than try to control it with hormonal products like the pill, my periods are definitely no longer the curse that they once were.

Now, I completely understand that if you've been on the pill or other hormone-based contraception for some time, the idea of coming off it may make you fretful. I totally understand that when faced with painful acne or other conditions that the pill is said to help with, the idea of NOT taking it is super scary. BUT you get direct feedback about your health from your menstrual cycles, so to suppress that knowledge can be counterintuitive to your wellness.

Personally, I try really hard not to put anything into my body that will change the way it naturally functions, especially if it's simply to control whether I get pregnant or not. I want to experience my cycle in its entirety, because when I do, I experience what it is to be *this* woman, in *her* fullness.

I invite you to do the same.

To experience your cycle in full.

To experience the ecstatic, orgasmic, anger, pleasure, and pain that occurs when you allow your true nature to express herself through your menstrual cycle.

To feel what it feels like to be in your body, experiencing the true expression of what it is to be YOU: someone who, in each phase of her cycle, embodies her power, potential, and potency.

Of course, there's no pressure to do *any* of this; your life is ALWAYS your call.

It's just that for so many years of being a woman, I walked around, zombie-like, totally disconnected from my body, my power, and my cycle. I used to thrash around in pain each month when my bleed came. I'd cry and scream, feeling so helpless and dissatisfied because I didn't trust my innate wisdom. I didn't trust it, because I didn't know about it, and now that I do, I feel a responsibility to share it. I want to make sure that every woman has access to all the information she needs to make informed decisions about her body, health, and wellbeing.

So if this is something you're feeling called to do, but are scared or worried about how to do it, I've got your back. I've asked my gorgeous friend, author, and reproductive fertility expert, **Maisie Hill**, to share her guide to coming off the pill:

THINGS TO CONSIDER WHEN COMING OFF THE PILL

So, you've been on the pill for so long that you've forgotten what having a menstrual cycle is like. Or, you remember all too clearly, thank-you-very-much, and the thought of returning to that hell is enough to bring you out in a cold sweat.

But times are changing.

You might be getting ready to start a family, or perhaps the pill is no longer managing the symptoms you originally took it for. Or maybe there's a small but insistent voice somewhere inside you gently nudging you to come off it.

Wherever you're at, I want to take a look at what happens when you come off the pill, and how you can take care of yourself as you make the transition.

Oh, and I'm not against the pill, by the way. I'm just a big fan of women having information about their birth control and reproductive health. The choices you make with that information is up to you… it's your womb, your business.

When you're ready to stop taking the pill, just stop taking it. You don't need to finish the pack you're on. Do bear in mind that you'll need to use another form of contraception as soon as you do because your body might decide to throw an impromptu "Woohoo, let's ovulate" party.

That may sound great if you want to have a kid, but there are good reasons to wait a few months before trying to conceive. Women on the pill tend to have lower levels of folic acid, zinc, and selenium, all of which play a vital role in early pregnancy as they help to prevent neural tube defects. So do your future offspring a favor by eating nutritious foods and taking a prenatal supplement.

Getting pregnant isn't an issue if you've no desire to actually have sex, but ladies, please don't beat yourself up if you fall into this category. Having a low (or non-existent) libido is a really common side effect of taking the pill, and unfortunately, it can take a while for it to come back once you stop taking it. Nourishing yourself by eating and sleeping well will help.

Your sexual desire may not be the only thing that seems to have disappeared, and if you're wondering where your period is, you're not alone. Most of the women I see when they come off the pill have either long or irregular cycles (a cycle being the period from the start of one bleed to the next).

Traditional Chinese Medicine (TCM) holds the view that taking the pill leads to two things: deficiency and stagnation. Think of it like this… you have a lipstick in front of you; there's not much left in the tube (deficiency), and the mechanism that twists it open is blocked (stagnation). Both of these things mean that nothing is coming out, or it takes a long time and a load of hassle to get some. Your period is the same because whether you want a baby or not, your body needs to believe it's strong enough to maintain a pregnancy to put the energy into menstruating. For ovulation and menstruation to occur, certain acupuncture channels need to be switched on, and the qi (energy) and blood within them need to be flowing freely.

When qi and blood aren't moving smoothly, women often get symptoms such as abdominal pain, bloating, tender breasts, mood swings, and irritability. The unpleasant truth from the TCM perspective is that years of being on the pill can actually cause PMS when you come off it. Essentially, PMS is caused by a blockage of qi. Don't go getting your knickers in a twist though, the good news is that acupuncture is pretty nifty when it comes to dealing with unpleasant PMS, irregular cycles, and other menstrual symptoms.

Smoothing the transition

There are several things you can do to make the transition a smooth one:

Improve your daily routine

If you skip breakfast, work long hours, eat (and drink) late at night, and don't get enough sleep, how can you expect your period to show up on time? When you're working on regulating your cycle, it's important to be regular in your routines. That means eating regularly (and enough), and going to sleep, or at least being horizontal, by 11 p.m. so your liver can cleanse your blood properly (and therefore clear the pill from your system).

Stop being an adrenalin junkie

Yes, I'm talking to you, Miss I-can-do-everything-and-I-can-do-it-faster-and-better-than-you. Slow down. Delegate. Your new motto is "Rest is Radical." Why? When you survive on adrenalin, you're really surviving, not thriving. Yes, we no longer live in caves, but our bodies haven't changed, so when your train is late, or you have a hectic day at work, your body responds by going into fight-or-flight mode. It's the modern-day equivalent of encountering a saber-toothed tiger.

When you're stressed, blood is diverted from your digestive and reproductive systems to vital organs, such as your heart and lungs, so that you've enough energy to leg it or stay and have a punch up. The consequence of this is that your reproductive system is regularly being disrupted. And there's more: your adrenal glands need progesterone to make adrenalin in response to stress, and the resultant reduced levels of progesterone can lead to PMS, infertility, and miscarriages. So there are lots of good reasons to take it easy.

Castor oil packs

Another affordable home treatment is a castor oil pack. Place the pack over the liver (roughly speaking, the section of ribs underneath your right breast) to help it to detox the pill. You can also place the pack on the lower abdomen if you experience menstrual pain or have any scar tissue from infections or surgeries that you want to soften.

Exercise

If things aren't flowing well, get in the flow! It doesn't matter if you're a fan of running, yoga, or dancing to Madonna in your kitchen, the key thing is to move in a way you enjoy, and which doesn't exhaust you. Remember, you want to nourish yourself, not cause further depletion.

Abdominal massage

Largely a self-care practice, this is a great option if you're restricted by money, time, or geographical location. A deep abdominal and back massage improves the flow of blood through the organs and tissues of the reproductive and digestive systems. When you see a practitioner for your treatment, they will also teach you how to perform the massage on yourself, so that you can then spend five minutes a day doing it. It's a great way for you to get in touch with parts of yourself that have possibly been untended for years, and the massage will see you through menopause and beyond.

So let's spread the word on this one. Let's create a REDvolution that says that women can confidently understand how their bodies work and take charge of their fertility for contraceptive and conception needs—without having to damage their bodies in the process.

CHAPTER 2

Menarche, Shame, and Menstrual Mysteries

My first experience of bleeding was less than celebratory. I was 13, and my mumma had walked out on my dad and I three months before. She left no address, just a phone number in case of emergencies.

I remember going to the bathroom and seeing blood on the tissue. I knew what it was because I remember Mumma having a supply of "jam rags," as she called them, hidden in her wardrobe for when she bled.

I felt really rather proud. I was bleeding like my mumma, and at that moment, I had a sudden urge to tell her about it. I welcomed having a reason to ring her, and I figured blood coming out of my vagina for the first time was definitely an emergency.

So I stuffed my knickers with tissues, went and found her number in the kitchen, and punched the numbers into the phone.

"Hello?" Hearing her voice made my heart yearn to be hugged by her. I loved my mumma so much. I doted on her and desperately wanted to make her proud of me.

I couldn't altogether fathom why she would have left me alone with my dad, who, if I'm honest, I barely knew, but I was sure she had her reasons. I was working on the basis that this was only a temporary arrangement, and when she was finished doing whatever she was doing, she'd come back and get me.

"Mumma, guess what? I've started my period!" I shouted down the phone. I don't quite know what I was expecting in return, but I was definitely expecting *something*. She was my mumma, and blood was coming out of my vagina. That was definitely mumma territory.

"What? Why have you rung just to tell me that?" She had said in a hushed, angry tone. "I said only to ring in an emergency! I've got to go; I'll be in touch."

Oh.

So many of us have less than stellar experiences of our first bleed, while others barely remember it at all. What I know now, that I didn't know then, is that by not marking this special occasion, by not celebrating the moment I became a woman, I remained a girl.

I see how so many of my actions in love, in relationships, in business, in life, have been informed by a 13-year-old me. In fact, I didn't fully come into my power as a woman until my 34th year, when, 21 years later, I had my very own menarche ceremony.

Not only was it incredibly beautiful—red carpet, rose petals, chocolate—I shared the experience with 16 sisters who were having their first menarche ceremony too. It was highly emotive and super potent.

The ceremony was three months after my mumma died, and a sister from the retreat took her role in the ceremony. She greeted me at the end of the red carpet covered in rose petals and lined with photos of our female lineage. She said all the words I'd wished my mumma was able to say to me on the day of my first bleed. She held me and gave me a gift that I'd prepared for myself earlier—a medicine bag filled with a shell from my favorite beach, a Mary Magdalene pendant, rose petals, and a love note to my 13-year-old self. From that moment, on that red carpet, celebrating my blood and my womanhood, something changed. If you've ever experienced any kind of ceremony or ritual, you'll know that you expect a subtle or maybe discrete change in your being—that's the power of ritual—but when stuff ACTUALLY changes, well, that can come as a bit of a surprise.

I walked taller.

I spoke more clearly and from a place deep within me. My words became womb-deep.

I was no longer avoiding womanhood; I now owned the woman I had become.

In many cultures around the world, a girl's transition to womanhood is celebrated in ritual and ceremony; it's the time of her coming into her creative and spiritual power. Yet so many women in the West have lacked this celebratory entry into womanhood, and it has affected everything from their attitude toward their menstrual cycle to how they view the body they currently reside in—and a million things in between.

Our menarche, our first bleed, is when our life purpose, our truth, is awakened in us. With each cycle, we speak more truthfully and nurture, respond to, and work with, the lessons it provides, allowing us to open and grow into the true awesomeness of who we are. **Imagine if we had known this at that moment of our first bleed—how would it have been different?**

For me, I'd have stopped trying so hard to be "someone" in my late teens and 20s—to achieve, to be liked, to be validated—because I'd have total trust that, with each bleed cycle, life was unfolding me, just as it should. There would have been flow, literally and figuratively, as I narrated the story of the woman that I am from a place of truth and purpose: my womb.

What's *your* menarche story?

What was *your* first bleed like? What did it feel like? Where were you? Was it celebratory? Was it negative? I invite you to put yourself in that young girl's body at that moment, right now and allow your heart to simply riff on it in your journal.

Depending on your experience, this may feel traumatic or a total non-event, but know your menarche story gives a powerful insight into the woman you are now, so dare yourself to *really* go there.

- If your experience *was* a celebratory one, how did that make you feel? How did it feel to have the moment you became a woman marked by family and friends? Were you proud? Were you embarrassed?

- If it *wasn't* celebratory, what was it like? How did it feel physically and emotionally? Where were you?

- Did you tell your parents? What was their reaction?

Revisiting your menarche story is like a deep-dive, self-excavation into who you are and why you do the things you do.

This is *your* story, and to become yourself fully, you need to claim it. Share it with your girlfriends, share it with your partner, share it with your children. Share it on your blog or in a sister circle in person or online.

If, once you've started to revisit your menarche story, you don't like what you've discovered, consider writing yourself a love letter in that moment of your first bleed. Tell that young girl everything she needs and wants to know at that moment—what didn't you hear that would really have helped you? Who would have spoken to you? Would they have hugged you? What would they have said to you?

This isn't a rewrite of your history; it's an opportunity to identify your feminine wounds and let the healing begin.

Menarche—the key to unlocking my story as a woman, by Michelle Hastie

Since starting my period around age 10, my relationship with it has always been love/hate (mostly hate). I loved feeling like a woman; a woman with curves, confident and sexy in my body. But I hated the cramps, heavy bleeds, the smell, the sensation of blood on my skin, and even the blood itself. Every 29 days for five days, I was holed up feeling awful and sorry for myself as my skin broke out, my breasts ached, and I could do little more than sleep.

I went on to have four children and an ectopic pregnancy all within seven years, and after that, my cycle became really irregular. I would bleed heavily for between two and six weeks at a time with clots and cramps. I barely had a week or two between each bleed. I was anemic, too, and everything the doctors prescribed failed to help. I became withdrawn and miserable while trying out all the various hormonal contraceptive pills and coils the doctors suggested. My health deteriorated, and I totally loathed my cycle because nothing worked, nothing gave me the balance I needed. I reached breaking point when I asked to have the Mirena coil removed. I no longer felt like myself; I had become so withdrawn and embarrassed by my skin and the weight I had put on. I was told by the doctor that there was nothing left for me to try. At 33, I was facing a hysterectomy.

Then I joined "Explore Your Lady Landscape," an online program run by Lisa Lister. I charted and followed the daily prompts, writing in my journal each day. I especially worked on the meditations and looked deeply at my menarche story. It was liberating. I had my first bleed at 10, and I knew vaguely about periods and sex, but missed the educational videos at school, so I was far from knowledgeable. When I saw the blood, I went to my mom and told her I thought I had started my period. She gave me a sanitary towel and a "facts of life" book. She seemed embarrassed, and it was clear from her behavior that she was uncomfortable talking about anything to do with periods or sex.

I read the book (later she told me she thought it was the best way because I have always been a reader) and it was factual. It told me a bit about everything, but it didn't tell me about the emotions involved, the tears that came or how I was supposed to exist with it. It was just a book of info. And so taking the lead from my mumma, I felt menstruation wasn't a subject to be discussed; I even hid my book in a very secluded space, pretending it wasn't there. My body changing seemed to bring about a lot of discomfort for my family. I remember my dad went into the bathroom, saw my bra, and freaked out. He told me that I shouldn't be leaving my "things" lying around for people to see. That it wasn't nice. That my body was changing, and I needed to have more respect.

From that moment, I became very private with my body and felt ashamed for bleeding and needing a bra.

I didn't realize I was carrying all of this until I wrote my story down. It was then I discovered how much I'd held in; how I had seen my bleed as a very private thing, rarely talked about it, and felt shame, like it should be covered up. But looking deeper, I've become much more comfortable with how natural it is, how much of a gift it truly is to be able to work with my own natural rhythms, and to honor that little girl who deserved more than a book to guide her.

Now I'm so much more open with talking about periods and am definitely passing this on to my children by being open and honest with them about their bodies and what to expect as they change and develop. I'm releasing the shame I felt as a girl growing into a woman, while holding a totally loving space for them to explore all the different facets of puberty and their developing bodies. Working with all of this has helped free me from the tension I've carried regarding my bleed. I can fully relax into it, without judging it, or judging my body.

Many women don't acknowledge or celebrate their first period, but it's never too late to press reset on your menstrual experience and throw yourself your very own menarche ceremony. In fact, it's the most beautiful gift you can give yourself. It's healing; if there's been pain and/or shame, it can create much-needed acceptance, and it's also an opportunity to buy or craft yourself a celebratory present, something that represents your bleed and signifies what it means to *you* to be a woman. It can be anything from a bottle of red wine to a deliciously decadent perfume to a beautiful bracelet or ring that you can wear every time you bleed. I host in-person and online menarche ceremonies, and my heart and arms are wide open to you if you'd like to initiate and celebrate the fact you bleed with me (this can be especially helpful if you're struggling with a menstrual health issue or are in search of life guidance), but creating your very own, unique-to-you, beautiful menarche ritual can be just as powerful.

It can be as low-key as a simple gratitude prayer to the Great Mumma under the light of a Full Moon, a lovingly prepared meal for yourself of your favorite food, painting your toes and fingernails scarlet red, or maybe inviting all the women in your family to join you for dinner and asking them to share stories of how they view you as a woman. Gather photos of the female lineage in your family and make an altar as you share the love letter you wrote to yourself out loud. Turn that love letter into a song to be sung with total abandon. Smear your menstrual blood onto a stone or crystal and commit it to the soil, either in your garden or land that you love.

Do what moves you.

But whatever you do, be sure to honor and celebrate you.

NOTE
....

If you just did a total cringe face at the idea of touching your menstrual blood, I dare you to get intimate with it. I dare you to touch it, to spend a bleed day without using any kind of sanitary protection, and just let the blood flow into a bath towel. If you're able, I invite you to sit on the ground and let your blood flow back to Mumma Earth, and while you're there, you could read Cunt: A Declaration of Independence *by Inga Muscio—such a powerful and insightful book.*

NOTE+1

If the word "cunt" and the idea of giving your blood to Mumma Earth freaks you out, you don't have to do any of this to get familiar with your cycle. But the easier it is for us to find ways to talk about our cunt/yoni/vagina/muff—insert your own word for her here—and the fact she bleeds, the less shame will be associated with her.

Truth

There was a time not so long ago that I really struggled with saying the word "cunt." It's generally used to shock or offend and can sound really harsh. But then I met the gorgeous **Colette Nolan**. Since I wrote this book, Colette has sadly left her body, and it was far too soon. Colette was a poet, artist, and Creatrix dedicated to celebrating "cunt" as one of the only words in the English language that women have to describe *all* of their genitalia. So, it's in deep bows to her knowledge and wisdom that I'm blessed that she shared with us the origins of the word "cunt" over the next few pages, and how we can cherish it and claim it:

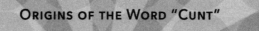

ORIGINS OF THE WORD "CUNT"

Originally, the "Cherish the Cunt" campaign was a personal healing mission called "I Heart My Cunt." It developed into "Cherish the Cunt" because we love men and wanted to move away from the outdated idea that all feminists are man-haters. We want EVERYONE to cherish women's cunts AND Mother Earth's cunt.

I knew resolutely that I would use the word cunt. It wasn't so much a decision but a feeling like I had no choice. There simply wasn't any other word that felt right. I've always liked the word cunt–coming from Ireland, we say it a lot more often than it's said elsewhere in the world. When I moved to England, I began to notice the shocked reaction that I'd get when I casually said it.

It was mainly from women, and this intrigued me. Language is so interesting with its loaded history and resonance; it's our thread to the past, yet it seems the reasons as to why the word cunt is considered such a bad word is no longer in our consciousness. When I've questioned people about why they find it offensive, they struggle to say why and proclaim, "It just is!"

If everyone knew the wonderful and varied origins of the word cunt, we would all be shouting from the treetops in celebration of the gateway to life itself. But people don't know the roots of this word, and for some unknown reason, the universe has plucked me, amongst others, to help them remember.

So first of all, cunt is the only word in the English language that we can use for the whole of the female genitalia. The word vagina, meaning tunnel or scabbard, a sheath for a sword, only refers to the inner entrance. Vulva refers to the clitoris, outer labia majora, and labia minora, but not the vagina. We need the word cunt if we want to talk about our incredible sexual orchestra in all its glory.

To do this in a liberated, empowered fashion, we need to educate ourselves about the varied and diverse origins of the word from all around the world. Of all the origins, the word cunctipotent *stood out for me. According to Barbara G. Walker, author of* The Woman's Encyclopedia of Myths and Secrets, *it means "all-powerful... having cunt-magic." Dr. Jane Caputi, a US professor of women's studies, has researched this word in-depth. She suggests that instead of "listening to our gut", as we are often told, we can move and speak from our cunt. "Cuntspeak," as she delightfully calls it.*

There are many varied and wonderful linguistic diversions I could take along the meandering path of the etymology of this explosive word. In order to keep it succinct, I'm sharing what I deem the juiciest findings, from goddesses and female figures in Africa, Sumeria, India, China, Japan, Korea, and Egypt all the way to the River Cunti (now known as the River Kennet) in Wiltshire, England and the stunning Síle Na Gigs in my hometown in Kildare, Ireland.

Africa

Kunta *means "woman" in several African languages. It has been found in ancient writings that there was a North African goddess called* Kunda Saharan, *and her tribe is still around today. They are called the Kunda. It is said that they can trace their roots right back to the cleft of the goddess. It is also recorded that Kunda Saharan was worshiped in the Saharan region before the area was a desert.*

This time was between 6,000BC and 8,000BC. The Kunda *people are now in Mozambique, Zambia, and Zimbabwe.*

Sumeria (Ancient Iraq)

Here the word kunta *means literally "one who has female genitalia." This is linked with the word* cuneiform, *which literally means "the sign of the cunta" or "queen who invented writing." Cuneiform is one of the earliest known forms of writing in Sumeria, dating from c.3100BC. At around the same time, there were priestesses named the Quadesha, who were accountants in the Temple of Inanna. It's highly likely that cuneiform was the form of writing the Quadesha used on clay tablets to record the temple's financial accounts, thus making it "the Sign of the Cunta."*

Inanna was the goddess of love, war, fertility, and lust. She was associated with the celestial planet, Venus. She was known as "Queen of Heaven", and the word qu *can also mean love, sensuality, sexuality, and the divinity present in all females. She was also connected with extramarital sex and sensual affairs, prowling streets and taverns for sexual adventure. There are hymns from Sumerian sacred texts glorifying Inanna's sexuality and singing praises to her beautiful and soft cunt. Interestingly, the Quadesha are also cited in some texts as "Sacred Whores."*

India

It's believed the word cunt came from the Proto-Germanic word kunto, *that's said to have come from the Indo-European word* kunti.

Kunti is the name of a much respected and revered Hindu goddess, also known as Cunti-Devi, and said to be the ruler of Kunta, which we know and call Kundalini energy, the snake-like feminine energy that travels up our spine. Legend has it that Kunti sang to the gods to call them to sleep with her. She eventually had a son with the Sun God, Surya, and the teachings of Queen Kunti can still be read today. Indian children who were born out of wedlock were known as Kuntas and revered as gifts of the Goddess Kunti. The word kunda or agni-kunda is also used in India for a hole or pit in the ground where fires are lit on altars in the Vedic religion.

China, Japan, and Korea

Remembering the cu, qu, and ku connection, we can understand how cunt is believed to be linked with the popular Buddhist goddess of compassion and mercy, Kuan-yin, Quan Yin, or Kunnon. She carries the Divine Mother aspect of Buddhism and is regarded by many as the protector of women and children. She's also revered as a fertility goddess capable of granting children. There is a Japanese school of thought that believes Cunda was the name of the Buddha's mother, but all I can find is that it was possibly the name of a female blacksmith who fed the Buddha his last meal of either mushrooms or pork. He fell violently ill and then became enlightened.

Ancient Canaan, Egypt

Here, Qudshu is venerated as the fertility goddess of sacred ecstasy and sexual pleasure, and she's depicted holding snakes in one *hand and a lotus flower in the other as symbols of creation. She's called "Mistress of All the Gods," "Lady of the Stars of Heaven," and "Great of Magic, Mistress of the Stars."*

All of this searching made me want to find links to Britain, where I live, and Ireland, where I'm from, and I was over the moon to find rich cunt roots on this side of the world too.

Britain

The River Kennet in Wiltshire was known as Cunnit until 1740, and many people believe this is related to our favorite word. It's a beautiful, vibrant river and a home to many species of plants, animals, and fish. One of the Kennet's sources is a chalk cave named Swallowhead Spring near Silbury Hill. This prehistoric artificial chalk mound is part of the UNESCO World Heritage Site that includes Stonehenge and Avebury. At 40m (130ft), Silbury is the tallest human-made mound in Europe and one of the largest in the world. It's thought to be approximately 4,750 years old. Many people believe that it was built as a representation of the pregnant belly of a Celtic goddess called Sil or Sulis, who was worshiped in that area of Celtic Britain and in northwestern France. British geographer and archaeologist Michael Dames believes that the quarry surrounding it was deliberately shaped to resemble the rest of her head, neck, and body. He takes this one step further by suggesting that if Silbury Hill is the pregnant belly, then the cave from which the Swallowhead Spring begins is the cunt with the River Cunnit flowing from it.

Ireland

There are ancient stone sculptures of women all over Ireland and the UK on old churches and castles. These incredible figures are known as the Síle Na Gig (pronounced Sheela Na Gee), and all sit in a squatting position with their hands holding their cunts open. Wide open. They are seen by many as "the divine hag" or "the sacred whore" or as a fertility symbol. It's believed that they were created to ward off evil, and the Christians allowed them to remain on churches to appease the Celts whose sacred sites they were building their churches on. I found two in my hometown, Kildare, recently and I can't describe the feeling of touching such ancient stone that was made with the intent to honor the power of women's cunts.

Cunt is used a lot in Ireland but not in the extreme and derogatory way it is in the UK and the US. We say it in an almost playful fashion, and we use the word cunteen, *which is a variation used to describe a young person. Legend has it that in Ireland, the word cunt was once a birthing call for women in labor. This one made me smile; I know what I'll be screaming when the day comes for me to give birth.*

"CUNT!"

Say cunt with a smile on your lips and love in your heart, and you will be surprised by how simultaneously soft and powerful you feel. That feeling is where we need to be as women in these tumultuous times.

You don't *have* to use the word cunt, but you should definitely let the word roll around your mouth. See how it feels to use it as a term of endearment for something that so many of us usually refer to as "down there" in hushed tones.

Our place of pleasure, our genitalia—our vulva, our clitoris, our vagina—really has been, until now, the area that shall not be named, and by not naming it, we're basically saying it doesn't exist.

Something so beautiful and powerful, yet the only names we can come up with are *pussy, mini ha-ha, flu-flu,* and *va-jay-jay.* Personally, I use *cunt, vulva*—because that describes the whole down-there situation—and I really like *lady landscape,* mainly because I love alliteration. I also use the Sanskrit term, *yoni,* pronounced *yo-nee,* which means *sacred womb space,* because I think it sounds sensual and love-filled—and that's how we *should* feel about this beautiful, sacred place in our body. If we don't name her something we love, how on earth can we show her love, or expect others to respect and honor her?

What do you call *your* lady landscape?

Really consider the words and phrases that you, and the people around you, use when talking about your vulva, your period, and your menstrual cycle. There's so much shame and blame associated with them that we've very limited language to describe them—least of all loving, sensual, and succulent language—and it's up to us to change that. Otherwise, advertisements that shame our yonis and vulvas—think feminine wipes and lip waxes—will continue to perpetuate the myth that there's something wrong with us and that we're not worthy.

Your vulva is perfect. It can be waxed, shaved, or plucked, or it can be left in its perfectly lovely, perfectly natural, untouched bush state. It's *your* property. I recommend respecting her, only sharing her on your terms, and pleasuring her regularly.

Use Colette's incredible insight to take you on a cunt hunt all of your own. Explore your own cultural beliefs regarding vulvas and bleeding, look at the words, language, and phrases that are currently used, pay close attention to the advertising of "feminine hygiene" products—how do they make you feel? If certain words make you cringe or recoil, dig around and find out why. Maybe your mumma didn't like it and passed that dislike down to you. Perhaps there's shame and embarrassment attached to a certain word or phrase—I was literally struck mute when I first heard the word cunt. I found it so vulgar and so harsh that the idea of using is at as a loving word wasn't an option for me. Still, as I got familiar with its origins, I began to practice saying it; I'd try it in different accents and, as Colette suggested, with a smile. I began to say it internally while lying in *Shavasana* at the end of a yoga class and holding my hands in a *Yoni Mudra*, and slowly, it has become a word I love.

Naming her, talking about her with love and affection, and honoring her as sacred will allow you to start trusting her as your guide, letting her inner wisdom move through your entire body. The same will happen when you embrace your cyclic nature. When you ritualize your menstrual cycle and live day-to-day more ecstatically, you'll become enriched by ritual and flow. As a result, you'll become more intuitive, vibrant, and satisfied.

You'll claim your worth as a woman, and, damn it, you are SO worthy.

Mary Magdalene and the Menstrual Mysteries

Wouldn't that be the best name for a girl band, ever?! About 15 years ago, a week after I was diagnosed with endometriosis, I found myself in Glastonbury, a small town in the southwest of the UK, which is world-renowned for being a place of pilgrimage for spiritual seekers. I'd never been before, but a weekend workshop to "connect with my inner goddess" appeared in my inbox, so I responded with an "I'm in, where do I pay?"

Little did I know that the weekend about to unfold was to be the start of a fierce, feminine, and devotional relationship with SHE. We were taken to temples and springs, and to sacred sites where we took part in rituals and songs. The experience was so far removed from my fast-paced life as a journalist—the job that was everything I thought I had ever wanted. Yet in those two days, I felt myself breathing more deeply, my heart opening wider, and a deep-down-in-my-belly yearning for this feminine connection that was being shown to me.

I mentioned this to Cordelia, our guide, who that afternoon, took us to Glastonbury Abbey, a place that has been soaked and drip-dried in mysticism and folklore. It's said that it's where King Arthur is buried, and it's supposedly where the Holy Grail is hidden—to name but two of the gazillion stories attached to this sacred site.

Basically, it's a pretty big spiritual deal.

We took part in a ceremony held at Mary Magdalene's altar, because, yep, she's said to have been there too, and despite my slightly cynical journalistic temperament at the time, I placed a rose for her on the slab of marble. I knew very little about Mary Magdalene at the time, but from how hard and fast my heart was beating and how hot my entire body had become at that moment, I knew that Mary and I were about to become acquainted. After the ceremony, Cordelia took us to the egg stone, a place not on the abbey map, where she said that women would come to sit and bleed during their menstruation. As my lady landscape and I weren't on speaking terms at the time, mainly because she was causing me crazy pain, the idea of sitting on a stone that celebrated the act of bleeding, didn't fill me with the love, joy, and giggles the other women in the workshop were experiencing. I watched as they each sat on the stone, having their pictures taken, while still trying to figure out what had happened at Mary Magdalene's altar.

When it was my turn, I sat on the stone, with my legs apart, and experienced the same heat I'd felt during the ceremony, the same fire I'd felt after leaving the hospital appointment with a handful of literature about hysterectomies rose through my womb, and into my entire being. My heart felt like it was going to burst wide open and shoot laser beams of light—kind of like a Care Bear stare, but a little less 1980s. It felt like an initiation of fire and light, which in retrospect, it was.

It was after this trip that my relationship with SHE

Who is SHE?

SHE is what I call innate feminine wisdom. It's the wisdom of our grandmothers and all the wise women that have gone before and all the wise women that are yet to come. It's the wisdom that's present and accessible to us each and every time that we bleed. SHE is every feminine archetype you know—bitch, witch, wise woman, priestess, virgin, queen, crone—as well as all those we are yet to discover. SHE is the goddess; SHE is Isis, Astara, Kali, Mary Magdalene, Joan Jett, Boudicca, Joan of Arc, Cleopatra, Taylor Swift, Maya Angelou, the Black Madonna, Síle Na Gig , Lalita Devi, and Akhilanda. This list? These are just a few of the ways that SHE shows herself to me.

SHE is your intuition. Your knowing. Your inner GPS system. Your power.

SHE power.

SHE is YOU.

YOU are SHE.

Mary is my homegirl

Since our first meeting in Glastonbury, Mary Magdalene has become my spiritual homegirl, my guide-ess of self-exploration on the path to loving myself and my lady landscape. And, after over 3,000+ years of lies and bad PR, I have been signed up as one of a team of 21st-century PR women to tell her *real* story. The story in which Mary Magdalene *was* the Holy Grail, the sacred container that carried the blood of Jesus: their child.

Yep, in some circles (mine included) it's believed that one of the most misunderstood and venerated women in history gave birth to a girl, and the truth, *their* truth, has been hidden from the world. And, quite frankly, they're both pissed.

We've been led to believe that Mary Magdalene was a whore (and that it's a bad thing), that women are "tainted," that our blood is dirty, and that sexuality and passion are evil.

I believe something different. I believe MM was an initiate of Isis, a priestess of the Temple of Isis, where women were trained in body wisdom, feminine arts, rhythmic intelligence, sex, and blood mysteries. In essence, these teachings are the inner mysteries of *all* women, mysteries that have been handed down from elder to daughter, and, well… the church got pretty scared of this kind of power.

So when they've told the story of MM, they've stripped her of her priestess powers and instead described her as a prostitute, a repentant sinner, and a wanton woman. (These terms and titles have been used throughout the ages to make us *all* feel guilt, shame, and blame for being sexual, sensual women with wants, needs, and desires.) In doing so, the church has stripped *us* all of *our* powers and created their own patriarchal narrative of what it means to be a woman.

When I first had the idea to write *Code Red*, I'd been studying the Divine Feminine for years and discoveries of ancient manuscripts were shining a light on Mary Magdalene as a powerful teacher and luminous feminine spirit. She wasn't just a disciple; she was JC's consort, his peer, his lover, and his teacher.

As I went deeper, I became obsessed with MM, the secrets of the Temple of Isis, and the ancient feminine blood mysteries—and became red with rage that the menstrual cycle had become hidden and shame-laden. What's more, the truth of MM, keeper of the menstrual mysteries, the scarlet woman (when I create my own makeup range, I am totally going to make a "Magdalene Red" nail polish and lip paint combo—blood red, obvz), has been both hidden and shame-laden too.

My anger and frustration were *her* anger and frustration.

The call to action to claim your power through the menstrual cycle in this book is *her* call to action to claim your power through the menstrual cycle.

One of the many things I love about Mary Magdalene (and there's SO much to love) is that she is a reminder.

A reminder that female forms of divinity can never really be repressed, because that which is feminine, will always show up, uninvited, unrequested, and unsolicited, whether you like it or not.

Mary Magdalene and the Menstrual Mysteries are calling to be reclaimed. To NOT tell her story, to NOT know yourself, your potency, and your power, to NOT claim the medicine of the menstrual cycle is to silence her even more, and, quite frankly, she deserves better. And so do we.

That's why you're here, that's why you're reading this book: you've been intimately summoned by MM—and damn, she's gathering quite the badass woman-gang right now—to rewrite the menstrual story. The menstrual story is the Magdalene story, and the Magdalene story is every woman's story.

I have been ovary-deep in research and delicious Magdalene and Sara Kali discoveries, piecing together their story for years now, and I'll probably continue indefinitely. I'm very much in it, and I'm without a full vocabulary to articulate my findings—but I'm embodying the teachings, the lessons, and the discoveries, allowing them to work through me and find their language for a modern time. What you're reading now is just the start. I want to tell the truth of the women that have gone before us and who are yet to come—time is not linear, remember?—because, in doing so, I tell our truth.

I want to do it in a way that's not just read about, but that can be felt, lived, ritualized, and fully experienced. Still, there was no way I could write this book without blowing a kiss of recognition to MM and the Menstrual Mysteries, and what's currently unfolding on my own rose-petal, blood-red path.

CHAPTER 3

Know Your Flow— Charting Your Cycle

We can reclaim the wisdom of the menstrual cycle by tuning in to the body's cyclic nature and celebrating it as our source of female power. Our bodies are deeply in tune with the cycles of the seasons, the elements, and the Moon. Unfortunately, we've become conditioned to multitask and take on the role of an all-singing, all-dancing woman of wonder. Someone who thinks resting is a sign of weakness and has a never-ending to-do list that... well, needs to be done.

By not rolling with our natural rhythms, more women than ever seem to feel subtle yet persistent anxiety, depression, and/or exhaustion. They're suffering from stress, burnout, and, more alarmingly, menstrual-related health problems. We're cultivating a dude-like existence—one where we need to control everything, strive hard, and do things in a linear, goal-orientated way. And this? Well, this results in us losing touch with our SHE power.

Our power to see, feel, and experience things that are beyond our control, things that our heart yearns for–intimacy, creative expression, authenticity, sisterhood. It's no wonder that we suffer and experience so much stress in modern life; our energies– physical, mental, emotional, spiritual, and creative–have been forced into a masculine structure, which simply doesn't fit with the natural flow of the female cycle.

The stories of the women that have gone before us have been edited, silenced, covered up, and censored, and it has meant that the patriarchal fear of our deeply feminine superpowers, has been stored in our bodies, in our DNA.

FYI
......

When I talk about the patriarchy, I'm NOT dude-bashing–I dig dudes A LOT. I'm talking about the systems that have been designed to keep us all compliant. Men are crying out for this knowledge, too; they want to step up to the Code Red *call to action. But they can't do it until we're able to really go there ourselves: to acknowledge and feel the rage and grief of Mary Magdalene, and to realize, like her, that we're not victims. We never were. The need to be beautiful and to fit a cookie-cutter version of perfection are all ways in which we are taught not to experience the dark side of our true nature. The side that shows itself during the second half of our cycle, the side where our deeply feminine truth and power resides–the messy, the wild, the untamed. Women have been shut up and shut down for generations, and, quite frankly, SHE, the Great Mumma, MM, the wise women, the ancient futurists of all genders–who have all carried this wisdom in their womb and in their blood–are suitably pissed.*

The time is now

It's time to connect with your body, to reclaim your power, to remember the truth of who you are. For us all to come together in sisterhood to forgive each other for competing and comparing, to forgive men (they didn't know, and it's up to us to educate them–not by fighting with them, belittling them, or trying to be them, but by being in our truth so that they can stand fully in theirs), and, most importantly, we need to forgive ourselves because we've done absolutely nothing wrong.

This long history of patriarchy has meant that we have an expectation of ourselves and of other women to be consistent, but the way of the feminine isn't linear. It's liminal. It's moment-to-moment, phase-to-phase. What I like one week, doesn't necessarily hold the same excitement or interest for me the week after, and nor should it. We all need to start getting really OK with being consistently inconsistent.

The menstrual cycle isn't just about your monthly bleed. Each phase holds information, emotions, and spiritual and creative insight. Yep, we have access to all kinds of incredible gifts, messages, and wisdom that can be used to enhance, amplify, and up-level our experience, and when we notice these changing energies and adapt our lives to come into alignment with them, we have an amazing opportunity to live and create in a way that honors both our rhythmic and cyclic nature.

For example, I wrote this book in complete sync with the gifts of my own menstrual cycle. I know that during my premenstrual phase I get truthful–some might say a little *too* truthful–so my writing becomes much more from the heart and the words flow much easier. I have access to a creative cosmic nudge that I don't experience so loudly, if at all, during other phases. It doesn't mean that I've only written my book during the premenstrual phase of each cycle, and it doesn't mean that I can't write during the other phases either, it just means that I've a monthly super-power that I can use in my work to help me create truthfully with ease. I would be a motherlovin' fool not to tap into that and use it, right?!

The ebb and flow of your dreams, your creativity, and your hormones in each phase of your cycle offers up an amazing opportunity to deepen your connection with your inner knowledge and power, and to live in a state of balance and sweet alignment with the different energies that show up for you each and every month. What would life be like if you knew that you could use your pre-ovulation phase to plan and schedule your workload for the entire month? Or if you knew that in ovulation, you're the Queen of communication? How about knowing you can use the premenstrual phase to get to the truth of the matter, to figure out what works and what doesn't, and how during menstruation, you can download directly from SHE/universe/source and let go and release what's no longer needed with your bleed? Cool, huh?

This is just the beginning. If you want to know everything that each phase has to offer, you're going to need to get geeky and start charting. Charting isn't just the territory of those who want to get pregnant; your cycle is a gold mine of incredible insight that will provide information on how best to use your energy, your intuition, and your power.

I've been charting my cycle for years now, and with each monthly menstrual cycle, I learn a little bit more about myself as a woman in the world. The more I learn, the more I'm able to know myself, and fully own my powers, care for myself, self-source, become more accepting of my body and my tendencies, articulate my experience, and live my life in ways that feel good and that are totally meaningful.

My relationship with my bleed cycle has, quite literally, brought me home.

When you work with your menstrual cycle and unlock this wisdom and the medicine that's held within it, you're setting out on a journey deep into yourself, one that you've been traveling since the beginning of time.

Many of us are told that our menstrual cycle is the five- to seven-day period of bleeding that occurs each month, and it's this bleedtime that's become known as our "time of the month"—but honestly? Your "time of the month?" IT'S. ALL. FREAKING. MONTH.

In fact, virtually every part of your body is affected during every stage of your menstrual cycle—pulse rate, blood pressure, body temperature, even the frequency of how often you need to pee. Your body is affected by it 365 days a year for as long as you have a menstrual cycle.

Now, just like the Moon, your menstrual cycle follows roughly a 29-day cycle, and in the same way that the Moon waxes, and wanes, so do we.

Each month, we take a journey through both the light and the dark parts of ourselves; we get the chance to renew and refresh our entire being—physically, mentally, psychologically, and spiritually.

In fact, in many indigenous communities, the words for the "Moon" and "menstruation" are interchangeable, because most societies have, at one time or another, understood the link between our cycle and the Moon. This book and the practices in it are *your* invitation to remember that you too know this to be true.

The Moon cycle

The Moon is the primary symbol for female energy. She takes about twenty-nine days to circle the Earth, roughly the same amount of time as the "average" woman's menstrual cycle. Both have energetic phases that hold medicine and intel that can support us as we navigate our experience as a woman. To understand our own cyclic nature, we can turn to the Moon and feel the energies of each of her phases on the collective. From new, the Moon grows and becomes bright and full; this half of the cycle is masculine, out-there energy. Then she begins to decrease in size and move back toward the darkness; this half of the cycle is feminine, inward-facing energy.

Waxing Moon

When the Moon is "waxing," she's getting larger in the sky, moving from the New Moon toward the Full Moon. This is a time of growth, a time to start new projects, meet new people, take risks, conceptualize ideas, to play and attract new love in your life.

Full Moon

When the Moon is full, she forms a perfect silvery sphere of gorgeousness in the sky. This is a time for those ideas you conceptualized during the Waxing Moon to manifest. It's a time to be seen. You may feel extra frisky and your senses may be heightened in this phase. At the Full Moon, feelings and emotions are amplified and illuminated—*everything* is seen—so if you're an introvert or you're hyper-sensitive, this phase can often feel a little "too much." What this phase does do, however, is let everything be felt and seen—so that you can use the Moon phases that follow to edit and release what's for keeps and what's no longer necessary.

Waning Moon

Once the Moon has reached its fullness, the Waning Moon decreases in size as it moves back toward the dark of the Moon. The energy of this phase is best used to edit, clear out, break bad habits or bad addictions, and end bad relationships. In this phase of the Moon, your intuition is heightened and your tolerance and patience levels may be lower than usual.

New Moon/Dark Moon

This is when the Moon is directly between the Earth and the Sun and is therefore hidden. This can feel quite intense, and you may feel the need to reschedule dates or plans in favor of alone time, meditation, and silence. The Dark Moon can also cause your feelings to feel like they're *over-feel-y*, so you might notice that you binge on food, TV, and social media to try to numb out. When the Moon turns new, however, set your intentions for the cycle ahead.

The wise women before us were in total sync with the Moon; they knew that our menstrual cycle is highly affected by the Moon's movement. Their blood and hormonal cycle followed the Moon's ebb and flow; as the cycle waxed from New Moon to Full Moon, estrogen levels increased, leading to supercharged fertility when the Moon was at her fullest, roundest, and most abundant. From Full Moon to New Moon, the waning half of the cycle, progesterone dominated and led to the release of blood at the Dark Moon.

Today, most of us are totally out of sync with the Moon's cycles. We get so busy in our daily lives that for the most part we ignore the cyclic changes of Mumma Nature—the seasons, the Moon, the ebb and flow of the sea—all of which are indicative as to how we *could* be living.

Our menstrual cycle is often medicated with synthetic hormones and not allowed to flow naturally from our body because we "stuff it" with a tampon. We treat any resulting pain and symptoms with

Western medicine, which means we're potentially silencing the messages that Mumma Nature, our body, and our cycle can give us each month. And when we do that? Shit will, and does, hit the fan.

We get stressed out, exhausted, depleted, and depressed.

We silence the deep, ancient feminine wisdom that's available to us through the cycles of the seasons, the Moon phases, our body, and our menstrual cycle by using caffeine, drugs, TV, and social media to create potentially destructive cycles of self-medication. These cycles keep us locked in the story that being a woman is hard freakin' work, that it has to be painful, and any alternative that involves resting or not striving to achieve makes us weak.

Now, I share this not to be judge-y or to make anyone feel "bad." We're all doing the best we can with the knowledge we have. It's simply a scene-setter so that we know where we're at. I'm also not sharing this to guilt anyone who doesn't bleed in sync with the Moon. In fact, one of the questions I get asked most is, "How do I sync my cycle to the Moon?" By which they usually mean: *How can I, like the wise women that went before me, ovulate with the Full Moon and bleed with the Dark Moon?* While this *is* possible, please know that it's NOT necessary.

A lot can affect our cycle, from blue light to street lights, to the amount of contraception in our water system, so it's rare that any of us actually bleed in sync with the Moon anymore. But despite what you might have been told, you really don't need to sync your cycle with the Moon—because guess what? It already is.

Whichever Moon phase you bleed in (and you'll notice this will change depending on what's being called up and through you to respond to, and work with, during the years that you bleed) holds its own power and magic for what you need to receive right now.

Here's *my* intel on what it means when you bleed during the various Moon phases:

Waxing Moon

You're being called to use your menstruation phase to explore, get curious, and make new discoveries about yourself and the world. Read books that hold teachings you've meant to read but haven't got around to or listen to podcasts that inspire you. This is a time to grow and play, taste different foods, take a left when you'd usually take a right. Dare to try new things and you'll care less about failing if you bleed during a Waxing Moon.

Full Moon

You're being called to share your work, medicine, and experience with the collective. Bleeding at the Full Moon is a call to use the vitality and potent power of the Full Moon along with the release of menstruation to bring something into being; to create, birth, and nourish outwardly, for the world. It will be no surprise that right now, many women will be bleeding with the Full Moon because they're being called to act and create in response to political and environmental situations. Their work here is to turn rage into creative action.

Waning Moon

You're being called to tend to the blooms and manifestations of the previous Full Moon. It's an invitation to use the information you've opened up to receiving during your bleedtime (you'll find out later, but you're a divination rod for Source when you bleed), to help you get geeky with the details, really stabilize and solidify already existing conditions, and verify the knowledge that will help to develop plans and projects.

Dark/New Moon

You're being called inward. To nourish yourself FULLY. Your dream time will be extra potent, so be sure to make lots of notes because when you bleed during this phase of the Moon, you have one foot in this world and one foot in the great void: the cosmic womb. You have access to all that's been and all that's to come—it's pretty wild. It's a good time to be still, to be in silence, and to let your wisest self be your guide so that you can set clear intentions for the cycle ahead.

How it is for you may be different to how it is for me, and that's OK. As you chart your cycle, take a look at the phase of the Moon and then turn your attention to how you feel. The menstrual cycle and Moon connection can hold a LOT of intel and wisdom for you.

In the same way that we can map our menstrual cycle using the phases of the Moon, we can use the seasons of Mumma Nature to deepen our understanding of our bleed cycle too. And there's no one better to take us on this particular journey than the queen of menstruality herself, **Alexandra Pope**. Alexandra is the co-author of the books *The Pill: Are You Sure It's for You* and *Wild Power*. She is a leadership coach, author, and teacher exploring the journey from menarche to menopause and beyond.

I've asked her to share how you can use the external seasons to map and understand the inner seasons of your menstrual cycle, and the importance of accessing the whole cycle each month—I LOVE THIS WOMAN!

THE SEASONS AND YOUR CYCLE

The phases of the cycle are rather like the seasons of the year: pre-ovulation is spring, ovulation is summer, premenstrual is fall, and menstruation is the winter of the cycle. Yet it's as if women haven't been able to experience the full order of the cycle. We've only been allowed the pre-ovulatory and ovulatory phases. The premenstrual and the menstrual sides of us are squashed, overlooked, and we have to wait until we get back to "normal" again. So, we only really ever fulfill the first part of the cycle.

Our pre-ovulatory time is our inner spring. This takes us out into the world again after the quiet time of menstruation. The power of the inner spring is a fresh beginning, full of natural focus and optimism, a sort of willingness to go out and try things again. It's a beautiful feeling; any cynicism or bitterness that we may have had as we came into the winter of our menstruation has melted away, and we feel, "Oh, maybe life is OK after all."

Then we come into our summer, a time of ovulation. This is when we have an easy, harmonious energy where we can be all things to all people, like a superwoman. Everyone loves you–it's a high-energy time and you really get stuff done; it's productive: you're magnetic and out there.

Your basic sense of self is strengthened, and your own needs are much quieter so you've got a lot of juice for others. However, if we spend our whole time tending to others and being "out there," feeling high all of the time, what will happen?

Women really get it when I ask that question. They say, "We'd be burned out, we'd be exhausted." I say, "That's right. And guess what? The wisdom of our cycle is so fabulous because the wheel turns." In the second half of our cycles, our bodies say, "That's great, you've done everything for everybody else, now you're making the trek back home to yourself."

After our fabulous inner summer, there's the agonizing moment where we cross into inner fall.

The fall phase of our cycle has many powers to it, and one is what I call "the great reality check." If you find this premenstrual time very difficult, it's because your psyche is telling you something; either something in your life isn't quite right, or you're doing too much for others. Don't expect to be loved in this part of the cycle as here as you may have a tendency to provoke, and if you don't use this energy consciously at this time, it will come out unconsciously in the form of PMS.

So, in fall, you start to turn in and become more tuned in to your inner feelings, which is why you can feel overloaded here. This increases the closer you get to menstruation, which is why it's important not to have too much stimulation going on.

Menstruation itself is our inner winter, and just as you do in the wintertime, you want to cozy up by the fire; the idea of partying sounds like no fun at all. At menstruation, we have that same impulse to retreat, to be quieter and still. Being aware that menstruation is a time for retreat makes a huge difference. Retreat is all about rest, repair, and renewal, and if we do this at menstruation, we can enter into the next new cycle refreshed and reborn.

Every month we have this incredible reality check with ourselves. Where do men get access to that? We're blessed. I don't want anyone ever again to condemn the menstrual cycle as a problem. It really is the most extraordinary self-care tool.

You see? Your menstrual cycle really IS the most extraordinary self-care tool, and this is just the start. Charting it, exploring it, getting to know it, beginning to build a relationship with your needs, wants, and tendencies in each part of the cycle is vital. On a super-practical, day-to-day basis, it means you're able to navigate life from a place of power, but on a spiritual and emotional level, it can crack you wide open to truth and self-discovery.

I was told categorically by a dude in a white coat whose attitude I didn't like one little bit, "You have severe endometriosis, you won't be able to have children." Up until that point, I hadn't thought a whole lot about children, to be honest. I was a career girl; I made money, wrote words, traveled, and had nice things.

Then, for a month or two in the sixth year of an eight-year relationship, I was pregnant.

I didn't want it.

That's not true. I *did* want it, but not with him.

I was preparing to tell my best friend—because I was totally going to tell her before I told him; my best friend is golden and ALWAYS knows what to do—when I felt a tug in my womb, a pain so strong and violent that it pulled me to the ground.

I was bleeding.

For the next three hours, I sat on the toilet and watched as my insides shed.

I cried.

I wailed.

And then for the rest of my life, until this very moment that I'm sitting here in front of the screen typing these words to you right now, I've never spoken about it.

Not to my best friend, not to him, not to my hot Viking husband, not to myself.

Such is the beauty of this work.

It WILL break you open.

It WILL peel back layers where you thought there were none.

After that bloody awful night, I went straight back to be the career girl with money, who wrote words, traveled, and had nice things.

I didn't honor that soul.

I didn't honor myself.

So, it was no surprise that, soon after, the relationship ended, and from that moment forward, life got a whole lot of suck.

I broke.

I broke so hard and lost everything: I lost my home; my career nose dived because I could no longer write the words I was being asked to write; my endometriosis hospitalized me. In fact, I lost any idea of who I actually was, and in writing this, I now see why.

I had to lose everything to find my way home to myself.

If I didn't break, I wouldn't have investigated endometriosis and gone on the adventure that has been an exploration of my lady landscape. I wouldn't have discovered the amazing power of my menstrual cycle. I wouldn't have trained with incredible teachers in menstrual cycle awareness, fertility, and womb wellness. I wouldn't be sharing what I know with you right now.

When you don't honor your cycle, your body, your womb, yourself—you will *always* break. Sometimes right away, sometimes a year or five later, but you *will* break because this is the way of the feminine. If you don't work *with* your cycle, you work yourself into a place of depletion—whether it's shooting for the top rung of the career ladder or keeping your shit together for the sake of your brand/family/parents/lover/children or any of the gazillion possibilities in between that women do in order NOT to break. Thing is, if you ignore your cycle, the monthly ebb and flow, the opportunity to create and let go, the opportunity that we are given EVERY month to give our body, mind, and spirit *exactly* what it needs in the phase that it needs it, you will break. This is because ignoring your cycle is ignoring SHE, the Divine Feminine. Basically, she'll keep encouraging you to break until, like the Hindu goddess Akhilanda, you become never NOT broken. I talk about her in most workshops, because when we as women realize we're never NOT broken, we stop using those broken pieces as a way to cut, shame, and punish ourselves for not being good enough, and instead see that the space between the beautifully broken shards is our place to learn and grow.

You know those moments when you feel useless and like you can't do anything right? Those raging body pains you experience during premenstruation? The need to receive outside validation to prove your self-worth? They are all your body's way of letting you know that something isn't cool, that you're suppressing an emotion or belief that isn't serving you, and unless you address it, it will come back bigger and worse each and every month 'til you do. That's how SHE rolls.

Now, just so you know, what I first thought was a breakdown was actually a breakthrough, because at 31 I met the Viking and my ovaries twitched once again.

After our first date, I texted my BFF the next day saying, *"I want to have his babies."*

Except I pressed the wrong button and accidentally sent it to him.

I know, right?

He messaged back: *"Good to know. When shall we start trying?"*

I love that dude.

On date three, I told him about the endometriosis and that I might not be able to have babies at all. I wanted to be upfront because if children *were* on his list of priorities, then I didn't want to give him false hope.

Turns out he *had* previously wanted children but said that he was so in love with me, he was happy for whatever was meant to be in our future together to unravel and unfold as the fates decide.

Yep, that's why I married him.

There have been a few times when my cycle has been 35 days, and we've both gone to the pharmacy for a pregnancy test, and there's been a mixture of both relief: *Phew, we love our life right now, thank you very much* and disappointment because *a li'l one would have been really pretty sweet too.*

I am totally of the belief, and this is because I'm a witch/hippy/ queen of positive thinking, that if a li'l soul chooses to come to hang with the Viking and me in this lifetime, then it shall be so. By knowing my cycle as intimately as I do now, by knowing what my body needs, what it's calling out for–movement, pleasure, hot sex, eight hours sleep a night, great books, Jo Malone perfume, love, a gluten-free diet (damn you, bread, why do you taste so good?!)–I can become a vessel of awesomeness, a fertile one of nourishment, should a li'l soul choose to come and hang with us too.

Now, if your monthly bleed marks the fact you're not pregnant, or perhaps the bleed isn't a period, and it marks the deep, gut-wrenching loss of a miscarriage, or maybe it's a reminder of a wanted or unwanted abortion–whatever your monthly bleed means to you, honor it, meet it face on, be in it. Don't try to suppress it, write it off as silly or irrelevant, or ignore it, because it will eventually break you. Allow your tears to flow just as the blood flows; feel into the darkness of what's shedding, just be there.

Scream, shout, sing, stamp, talk, hum, cry.

Expect nothing from yourself except whatever emotion, feeling, sensation shows up.

Find a place in nature that you can go to when you feel like this. Mine is the sea. Saltwater is my healing balm.

As I write this, tears are flowing down my cheeks at the realization that a soul *did* choose me, if only for a fleeting time, and maybe there was a reason for that.

If they hadn't, I wouldn't have broken. I wouldn't have left the dude I thought I was meant to spend forever with. I would have been much less likely to find a life path that wasn't career orientated and where I cared much less for material things. And most importantly, I wouldn't have met the Viking.

My soul mate.

My beloved.

So riff the shit out of your experience.

If I hadn't let my heart riff this morning over a cup of licorice tea with you, that particular part of my story would have remained a locked box and who knows what damage that would have done.

How to chart your cycle

I invite you to read this book from cover to cover, then for the next three months at the very least, I'm asking you to commit to charting your cycle. (I've created free downloadable cycle charts online, see REDsources at the back of the book for the link and password.)

Now to get an idea of how this rolls, please refer to the filled-in, sample bleed cycle wheel—you can download that too. It's from my very own cycle journal so you can get a sneaky peek into what makes me tick *and* get some guidance as to how to explore the rhythms of your own cycle too.

On the chart itself, I've also included diagrams of a 29-day moon cycle to show its different phases throughout one lunar cycle, or one menstrual cycle.

The black moon indicates the Dark/New Moon, and the white moon indicates the Full Moon. The moons that appear in between the New and the Full Moon shows the moon either waxing to full or waning back to new.

I'd recommend you start to chart on the first day that you bleed.

This is day 1 of your cycle.

Now find out which phase the moon is in by looking up the date you begin bleeding on a lunar calendar—you can Google "lunar calendar" to find out exactly what the moon is doing from one day to the next or download an app on your phone. For example, if you start to bleed on January 4, check the calendar to find out the phase of the moon on that day.

Record the days and the date.

Chart your emotions: how you're feeling physically, mentally, and spiritually—make a code/legend if there's not enough space. If your geek-girl-self starts to fall in love with this process, start journaling it too.

If you journal already, add this charting to your practice. If you're called, you could also start charting the astrological journey too, as this also works on a cyclic basis. Throughout our 29-day moon and bleed cycles, the moon passes through each of the astrological signs for a short time. So, you might find that while you don't bleed exactly every 28/29 days, you might be bleeding every time the moon passes through Virgo, and then you can go look up why Virgo is so significant in your life right now—seriously, the self-discovery is NEVER ending.

I LOVE it.

> ### FYI
> ....
> *If all this sounds a li'l technical and complicated, don't fret. I've made a super-short video to help make it as easy and as pain-free as possible which you can download along with the charts (see REDsources at the back of the book for the link and password).*

Why chart your cycle?

Your cycle is the key to unlocking what's good and what's not so good in life. So if you're suffering from PMT, it's your body's signal that you may not be paying attention in the other phases of your cycle. Maybe you're working too hard, or not taking time out for yourself. When you start charting your moods and actions for a few months, you'll begin to see patterns emerge, tendencies will make themselves known, and you'll really get to know *your* cyclic nature.

You'll see which days in your cycle mean you want to curl up in a ball and hide under the covers and the days when you're so full of energy you could work for 14 hours straight and still take on the world (whether you should or not is a different matter!) Once you've read this book and in particular, the following four chapters that highlight each phase of the cycle in detail, you'll be able to start starring the days when you're really rocking out the days when you're teary, and the days when you want to gouge people's eyes out with a spoon. One of my most favorite realizations-through-charting was from a client who noticed that on day 27 of her cycle, a few days before her bleed each month, her husband "drank water like he was chewing carrots." Best. Quote. Ever.

Start to see when your family really annoy you, when your lover is so irresistible you text them at work saying: *"Get home now, I want you,"* and when you actually want to do housework (I hate housework, but there's one day, each cycle, day 21 for me, when I actually just want everything to be put away and clean— it was a revelation!)

When you start to pay attention to what comes up for you each day, over three months, you'll see patterns begin to emerge, patterns that will become your personal hot spots and superpowers. I'll be giving you an overview of each phase, but when you start to chart, you start to get super-specific to what's relevant to you, and you then begin to create your very own menstrual map—because forewarned is forearmed, right?

As you become more aware of each phase of the cycle and how you personally experience each one of them, you'll find it much easier to recognize and use the superpowers of that phase. You will also experience a far greater acceptance of your body, your menstrual cycle, *and* your complete feminine nature.

Your menstrual cycle is *way* more than just a biological process; it's a cycle of ever-changing spiritual, emotional, creative energy, a road map that leads right back to the very essence of you.

EVERY MONTH.

When you chart, you'll start to see that you experience life through the lens of the current phase that you're in. How you see the world will change from week to week, so at each phase of your menstrual cycle, you show up to life differently. How I answer a question during ovulation will be *very* different to how I answer a question at menstruation. Can you see how knowing this information could actually start to really help you plan dates, plan business meetings, plan your entire bloody life? And, more importantly than all that, it will help you to realize that you're *not* crazy and that you're *not* consistent: you're a woman, you're cyclic, and that's NOT a bad thing.

In fact, just so you know, it's a really bloody amazing thing.

Being consistent is a masculine construct and doesn't altogether serve us as women. When we try to work, create, and love from this place, we end up exhausted and depleted because it's simply not sustainable for us.

You CAN do it, but at some point, as I mentioned earlier, you *will* break. It makes a LOT more sense to stop trying to impose a false model upon women which suggests we always need to be seen, wearing a game face and taking action in order to validate ourselves as "successful."

The magic of the menstrual cycle means that every month, you have the opportunity to nurture and honor yourself, to renew and restore your energies—not just physically from the shedding and renewing of your eggs and womb lining, but your confidence, your dreams, your sexuality are *all* renewed during menstruation, ready to be used again in pre-ovulation. How freaking awesome is that? And the good news is that you'll find you can do *and* accomplish more without feeling useless, exhausted, or frustrated. In fact, you will maximize your impact, your influence, and, more importantly, your pleasure in the process. Hurr-freakin'-rahhh!

It's why I'm OBSESSED with sharing this ancient-yet-futuristic, cyclic, menstrual wisdom and knowledge and making it totally relevant and accessible for modern women like you and me.

Now, what I found when studying this work is that a LOT of it is deeply seated in either the very practical, scientific biology class-style teaching or the hippy-dippy, super-spiritual love-your-womb offerings—I've studied and love on both. But I REALLY want everyone to know their flow, unlock their monthly superpowers, and create a bloody amazing life because so much of our suffering is because we've lost our connection to the power that is locked in the menstrual cycle. I want to help you crack your lady landscape code and restore this knowledge in the most simple way.

What I'm going to share with you over the next four chapters is a combination of my personal discoveries from my own menstrual experience, and the wisdom that has occurred from sitting in a circle with women, both in person and in my online programs.

What you'll see and feel as you read through each chapter is that there's a common thread, a blood-red archetypal strand that connects us all–but we will *all* experience our cycle differently. As you get to know each phase intimately, you may realize that you totally love the newness and ability to get stuff done in pre-ovulation, but you hate ovulation as it feels too bright and too "out there." Or maybe you feel energized during your bleed which feels totally different to how others experience it–your job is to pay attention to all these discoveries, to chart them and to journal them as they unfold themselves to you. They hold important clues about areas that may need development or attending to on the ever-evolving carousel of being YOU!

SHE STORY

The joys of charting, by Maree Gecks

Before I started charting, I couldn't work out why I felt so completely different from one day to the next. Not just a little bit out of the ordinary, but a totally different person with different perspectives, ambitions, needs, and feelings. I know I'm a Gemini, but we're not talking twins here—we're talking quadruplets! Even my friends would comment, "Which Maree are we going to get today then?!" I was beginning to think that I might be going mad! Why couldn't I be consistent and straight-talking like a lot of my male friends? I'd been following Lisa's work about the ancient wisdom of women and the power we harness, so when the Chart Your Cycle course landed in my inbox one morning, I knew it was something that my business and I could definitely benefit from.

When I first started charting, I didn't even know what day 1 of my cycle was! Was it the day I started bleeding, the day I stopped bleeding, or just the first day of the month? I really was clueless about my menstrual cycle, which was pretty worrying, considering I'd been having one for 16 years.

I knew I got pretty teary and grumpy for no apparent reason just before my period, but it would only be when I started bleeding that I would get that *Ahhhhh—that's why I've been bursting into tears when a customer sends me a lovely thank you letter* realization. I could never understand how one day I would be full of energy, fearless and ready to take on the world, and the next day I would want to stay in bed and be still, quiet, and alone. I really was beginning to think that there was something wrong with me. Thankfully, through charting a wonderful web of me and all my different personas, strengths, and weaknesses began to unfold. I'm not going to lie, the first month was hard—just trying to work out what I was feeling and putting it into words was a challenge for me. I'd never really connected with myself on a daily basis like that before. But the more I did it, the easier it got, and I quickly started to notice patterns.

I stuck with it, and I'm really glad I did because now I have a deep appreciation as to how to work with my cycle and not against it.

When I'm bleeding, I'm super connected to my spiritual side.

Meditations are so powerful, a slow yoga class is luxurious—I love being alone and I really feel like I can tap into my female intuition. This is a perfect time for me to take stock of where I am in life and in my business, and to really feel what is and isn't working for me. I've come to some pretty big realizations during this time, by just sitting and listening. When I arrive at pre-ovulation and ovulation, however, I'm a totally different person. Sitting in meditation is really hard as I'm literally brimming with excitement, energy, confidence, and ideas. I want to know everyone and everything, and I want to be everywhere all at the same time. This is a great time to meet new yoga teachers, to make sales calls, catch up with existing stockists, film videos for my blog, and concentrate on social media.

My premenstrual phase is my least favorite. It's when the doubt starts to kick in, and it's when my energy levels plummet. Rather than being afraid, confused, and fighting against this change, I now know what's happening in my body and what my body needs. It's a great time to rest, nourish, and recharge. It's when I decide to work from home, have an early night, catch up on research, and finish those boring jobs like accounts. When I give myself permission to do this, not only does it feel empowering, but it also means that the month ahead is so much easier.

I've definitely still got a long way to go and so much more to learn, but you know what? It's exciting finding out about yourself, especially when it makes such a difference to your life and the way you work. It was a complete revelation to me to understand that I was a cyclic being and that trying to be consistent and the same person day in and day out just wasn't who I was as a woman. It's not how I was designed, it's not how I work, and it's just not me!

I've now started to look at my cycle as a blessing; something to be celebrated, something that makes me special, rather than something to be fought against, ignored, and changed. While I felt huge gratitude for Lisa's teaching, I also felt angry that I hadn't been taught this earlier. I'd spent nearly three decades trying to be something that, by nature and by birth, I was not.

I urge you to give it a go. It's natural to be a bit skeptical and to write it off as "hippy-dippy," but I promise you, it's life-changing. It really has transformed the way I view my cycle, myself, and being a woman.

Before you dive into unlocking your lady landscape code, I have one caveat.

There are NO rules regarding the cycle.

While bleeding at the Dark Moon and ovulating at the Full Moon is something that happens for some women, others may bleed during a Waxing or Waning Moon. Some women bleed for three days, while others bleed for eight. There are no rules. I repeat, NO rules.

While I may suggest that ovulation is a time to be sociable, once you begin to chart, you may realize that being sociable at ovulation is the last thing you want to do.

I really dig on my premenstrual phase, but you may find big love for pre-ovulation.

It's **ALL** good.

Everything I share here is in broad brushstrokes and rough outlines; the details and the coloring in are all down to you.

This is *your* exploration, *your* journey of self-discovery. Reading *Code Red* is definitely a start, because simply knowing this information *will* start to change how you think about the cyclic nature of your life, but if you *really* want to experience the full awesomeness of your cycle, you have to do the work. And it *is* work. Sometimes, as I explained earlier, it will break you open, sometimes it will be challenging, sometimes you'll want to sack it off and wish you'd never even picked up this book, while other times you'll be toasting Mary Magdalene with a glass of red wine in total appreciation for the self-discoveries that have unfurled for you.

This unfolding and monthly exploration of YOU can be as spiritual and/or as practical as you need it to be. It's completely up to you whether you use it simply to find your superpowers (the days when you'll ace a presentation, the best day to ask for a raise, the days to have an argument and know that you'll win, the days when you'll want to go on top, the days when you need to work from home because there's a chance you may say something you'll regret—oh yes, they're all in here) or whether your cycle becomes your spiritual practice as it is for me.

I used to be a total self-help junkie. My bookshelves were full-to-the-brim with variation-on-the-same-theme book titles. *How to be better/do more/lose weight/gain pounds in 21/30/60/90 days*— you know the ones, right?

Then, as I got to know my flow, I realized that while all these books had their place, I didn't need them (a discovery my accountant is very happy about, what with me previously spending more money on books a year than I do on household essentials like food and clothes, but I digress). Nope, self-development is not the goal. The goal is self-discovery, self-knowing, and self-actualization, and your cycle is an all-in-one self-care/spiritual/organizational/relationship tool that not only can be used as a super-practical daily life planner; it can also be used as a deeply sacred daily devotional to yourself.

It really is your call.

The synchronicities of the cycle, by Nikki Turner

I'm shocked at the synchronicities, the insights, and the self-knowledge I have gained from charting my cycle.

Where once I would completely go against my true nature, I'm now in alignment with it. When I'm arranging my diary, I know when is the best time for certain appointments, events, catch-ups, business launchings, when is a good time for "me" time along with the *seriously do they not know it's day 24?!* time.

Even when something crops up that isn't in alignment with my cycle, I now know what I need to do to prep for that and to honor myself with self-care and compassion. In truth, I know myself on a deeper and fuller level than I could have ever imagined. I accept myself and my natural cycles. I love them. There. As cheesy as it sounds, I said it. I love my cycle and the gifts it gives me. It's so amazing that my body is in tune with the world around me. Through charting, I have seen a connection to the Moon cycles, to the seasons of the year. Everywhere I turn, I see cycles that reflect my own back to me. I truly believe that THIS work, this awareness and understanding, NEEDS to be known by women, children, men, employers, employees… everyone. For if we did know this, if we were aware, then the power of women could be seen for all its transformational nature.

What you need to get started

Download your charts–Go online to find sample charts from my cycle journal, print-off-able and fill-in-able charts, and a marvelous menstrual mandala to print off as a guide. Find the link and password in REDsources at the back of the book.

Make a playlist–I have taken props from the movie *No Strings Attached* and created a Period Playlist to accompany this book. If you don't dig my tunes, make one of your own–let your mood lead you to the songs that move you and soothe you in each phase. Then please share it and hashtag it on social media with #coderedthebook–I LOVE a playlist.

Make a menstrual medicine box/basket–A beautiful way to acknowledge and make our cycle sacred is to create a box/basket/shelf/bag of menstrual medicine.

This can be as basic or as decadent as you like. I always lean toward the decadent because that's how this particular period priestess rolls, but creating your own menstrual medicine box, that you're able to pull out a few days before your bleed, will act as a reminder that this is a time to listen to your body and nourish yourself.

- **Journal**–I never need any excuse to buy stationery, but as you'll find out later, during your bleed your right and left-brain hemispheres merge, and you have seer-like insight into what works and doesn't work in your life. Make this journal beautiful and sacred. I have a special inky red pen that I keep with it too.

- **Oracle cards**—I've created both my oracle decks—*The SASSY SHE Oracle* and *The SHE Sirens Oracle*—specifically for working with the whole of your Moon/menstrual cycle; both decks are the many faces of the feminine and aspects of what it is to be a woman. I use them as guidance in each phase, and the gypsy-witch tarot reader in me also created a spread dedicated to providing guidance during your cycle, so these cards are supercharged to provide menstrual medicine. But you can choose any cards that resonate with you.

- **Essential oils**—I've highlighted essential oils and sweet-smelling perfumes that may benefit you in each phase of your cycle, but really if you have a favorite scent or smell, or a particular need or requirement at bleedtime, pop the essential oil that's right for you in your basket. A heart-opener like violet is great, as is vanilla which can be calming when we're in the darkness of our cycle.

- **Crystals**—Not only are crystals pretty and lovely to hold, but they also have deeply healing properties. Again, there are specific ones that I share throughout the following chapters that will be beneficial in a certain phase of your cycle, but if you're new to working with crystals, simply choose one that you're called to and then look up its meaning. There's a chance that it will provide you with the medicine you need at that time. I have a carnelian palm stone in my box that has specific healing properties for reproductive organs.

- **A red piece of jewelry/scarf/hair accessory**—This is something beautiful that you wear only when you bleed to signify that you're in your power.

- **Incense**—I love to burn Lady Nada incense when I bleed, and it's now become a smell that I associate with my power and potency. If the idea of burning incense sticks doesn't roll with you, pick a perfume that you wear only at bleedtime. I've shared my favorite perfume for each phase in the following chapters—I wear Hugo Boss Red when I bleed: it's musky, and it's powerful. Smell can really help us to ground down into our root chakra, which is why I like the idea of having a different smell and color association for each phase—so I can mark each different energy change as I move in and through it.

- **Chocolate**—Preferably raw or dark chocolate really is medicine, and the goddess of chocolate, Ixcacao, is a powerful deity to call on during your bleedtime. (See also REDsources at the back of the book.)

I also have a goddess statue ritualized with my menstrual blood, runes, and a massive conch shell. I have a MAC "Russian Red" nail varnish that I only apply at bleedtime. Some of the women I work with have actually started to ritualize the entire cycle, finding what their needs are in each phase and creating a box for each season. I bloody LOVE this idea!

Ohh, there's one last thing. I mentioned it earlier, but I'll remind you again here: if you're on the pill, pregnant, don't bleed, or no longer cycle due to menopause or surgically induced menopause, this doesn't mean that you miss out on the cycle action, insight, and wisdom. While your body won't be experiencing a menstrual bleed, it *will* still experience its own cycles, but these may be much more subtle. So to fully embrace the cyclic energy, I'd recommend working with the phases of the Moon as we cycle through the phases in this book—I've highlighted which cycle phase represents which Moon phase at the beginning of each chapter. All the traits, archetypes, and superpowers are the same; it's just that through your menstrual cycle, you experience these through your body and through your blood, making the practice of cycling—no bike required—totally unique to you.

Here's how we will proceed

I'll give you an overview of each phase, what it feels like when you connect with it with all your big beat-y heart, and what can happen when it's ignored. Then I'll suggest a myriad of ways to help make the most out of each phase, connect with its SASSY—**Spiritual, Authentic, Sensual, Sensational YOU**—superpowers, and become more conscious of your personal power source. At the end of each chapter, the very lovely **Em Tivey**, a herbalist, and fellow menstrual maven, shares herbs and tinctures that will benefit you most in that particular phase, and one of my most favorite people **Salvatore Lomonaco**, maker of Floralunity flower essences for sensitive and intuitive souls, shares his flower medicine to help you connect with the floral essence of that phase.

What really matters is that you simply use the information I've shared to start following your own inner guidance as you travel through each menstrual cycle in a way that feels real and relevant to you.

Use this guide as a way to get familiar with each phase—some women want to slow down or totally stop during menstruation, while some women have huge ecstatic visionary bursts of ideas, almost like a kundalini awakening. How it looks for you will look totally different to someone else, so start to explore how it awakens and manifests within you and then please share and tell me all about it!

CHAPTER 4

Pre-ovulation–Like a Virgin

Female archetype: Maiden

Moon phase: Waxing Moon

Season: Spring

Element: Air

Color: Green

Tarot card: The Fool

Crystals: Citrine—energizes and cleanses

Adventurine—inspires optimism and a zest for life

Essential oils: Jasmine—inspiration, passion, and joy

Lime—stimulates and is great for purification

Mantra: "I get shit done."

Song: *Like a Virgin*—Madonna or *Just a Girl*—No Doubt

Perfume: Daisy by Marc Jacobs

The science bit

Pre-ovulation is the first phase of your cycle and usually begins around day 7 all the way through to day 13. (Obviously, this will vary depending on the length of your cycle, and it may take a few months of charting to really start to see where each phase begins and ends for you, but that's part of your very own lady landscape exploration—you need to get geeky, remember?!)

So, in the 7-10 days after your bleed, you'll start to come out of your winter-like cocoon. The steady increase in estrogen boosts your brain's serotonin levels, which leads to an increase in energy and enthusiasm for… well, just about everything, and you'll feel a lot more upbeat than you did in your previous bleed days. Hurrah. You may want to spend time with girlfriends, dance, start a new class, or learn a new skill. Words come easy, and you're articulate, so if you've got a big presentation or an important call to make, definitely do it in this phase of your cycle.

Like a virgin

Pre-ovulation is like the Maiden—fresh, new, young, and full of hope, possibility, and potential. It's dynamic, active, and radiant. In fact, when I did a visualization with a client to take her on a journey to meet her SHE guide-ess in each phase of her cycle, because of the very nature of who she is, she visualized each guide as a movie star, and in pre-ovulation her guide-ess appeared as Dorothy from *The Wizard of Oz*. Dorothy just seems like the perfect encapsulation of this virginal, pre-ovulation spring phase. She finds herself stepping onto a yellow brick road in this incredible new world of technicolor—she's naïve, wide-eyed, and a little trepidatious, but the possibility of what's to come is palpable as she sets off on her new adventure. This is exactly what's happening to us each and every month. Like the Fool in the tarot, you're wide-eyed, fearless, and full of hope for what is to come. With each cycle, your pre-ovulation is a brand-new page, an opportunity to write a new chapter in the ever-unfolding story of you. Like Dorothy, you have the opportunity to set off on an adventure that will always lead you back home to yourself. Except, unlike Dorothy, your path isn't yellow, it's blood red. (And, of course, the ruby red shoes are optional, but come on, who wouldn't want a pair of sparkly ruby red shoes?!)

In a culture where we've been taught to people-please and to put the needs of others before ourselves, the Maiden, with her carefree attitude to life and openly flirty, sexual expression, is often frowned on. Yet her youth and vitality are what's held up by society as the beauty ideal, so you can see why there might be some confusion. Much like the confusion you feel at puberty when you've got all those sexual feelings, and new hair and boobs are appearing on your body, yet you're still hankering to play with dolls and get a hug from your mumma. To quote Britney (and why would you NOT?), it's that whole *"I'm not a girl, not yet a woman"* thing. But this is why when we come into our pre-ovulation phase each month, despite what puberty was like for us, we feel a sense of ease, because this is known territory for us all. However, if during your own Maiden years you weren't crowned prom queen, or you didn't have the swishy just-stepped-out-of-a-salon hair, or maybe your first sexual experience was less than ideal, or your first bleed holds a lot of shame, there may be some sub-conscious angst or pain that makes this particular phase less than OK for you. I invite you to explore that. To see if there's anything from your own Maiden phase—for some of you, there's a chance you're still in it, and for others, it may be some years since you were in the Maiden territory—that's holding you back. Don't think too deeply: just get still with yourself, set a timer for 15 minutes, and journal your heart out about what first comes up when you think about puberty, your first bleed, your first kiss, your first sexual experience, your appearance at that time. These are all golden nuggets that will help you to understand this phase of your cycle, how it currently shows up for you, and develop a better understanding of yourself.

I did say there'd be work to do, and it wouldn't all be easy, didn't I? What I can absolutely promise you though is this: it CAN change your life.

Think Athena—the strong, self-defined goddess—highly intelligent, rational, outgoing, practical, and logical; or Artemis, truly present and reveling in her body, pushing it to its physical limits.

These Maiden traits are all available to you in your pre-ovulation phase.

Think of it like spring or the pagan Sabbat of Imbolc: after the death and darkness of the winter, we're all given a chance for a brand-new beginning, to start afresh, to plant new seeds and to watch them grow.

Your pre-ovulation SHE powers

Each phase has a set of superpowers that you can access and hone to make life a whole lot sweeter during this part of your cycle:

- At pre-ovulation, you have memory, logic, and reasoning. Hurrah. These help you to understand and make sense of both people and projects fully, and, more importantly, you also have the drive and determination to manifest your ideas.

- Your physical energy is rising in this phase, which means your stamina is renewed. You're active and able to fully engage in life: you can rock out on the dance floor, in the gym, and most definitely in the bedroom. Wink.

- You're an idea generator in this phase, and you've the physical and mental energy to work for longer to get shit done. If you're going to pull an all-nighter, *this* is the phase to do it in.

- You're more available for social engagements and are more able to be seen and to voice your opinion.

- Fearlessness, self-confidence, and self-belief are more accessible in this phase than in any other.

During pre-ovulation, your energy can feel upfront, self-assured, and high-spirited. It's masculine, it's yang, and this is the energy of movement, of being in action. It's activated, engaged, and has a neon light above it that flashes: *Watch out world, here I come.*

How to activate your pre-ovulation SHE powers

You've access to these particular SHE powers during the pre-ovulation phase of your menstrual cycle EVERY. SINGLE. MONTH.

The best way I've found to fully activate them and use them to their full potential is by aligning your SASSY–Spiritual, Authentic, Sensual, Sensational YOU–but remember, in each phase, it's up to *you* how you choose to use and interpret the information. This is just simply a guide to help you on your way, OK?

Spiritual

In pre-ovulation, there's likely to be a desire to try out lots of different approaches to your spiritual practice to see how they "fit."

You might want to sign up for a meditation class, embodied art, 5Rhythms, AND hot yoga on the same day. You'll potentially feel like a wide-eyed child wanting to try everything all at once. Be open to where this exploration takes you, don't be afraid to experiment. If you'd never considered going to see a psychic, yet your BFF has two tickets to go during your pre-ovulation phase, go. You may be suitably surprised.

If you already have a strong spiritual practice, you may feel called to express it more outwardly. So whereas you may have sat silently in front of an altar and prayed to Maa Gayatri 108 times with mantras and malas during premenstruation, you may feel called to express your devotion a little differently in pre-ovulation. Maybe you want to do sun salutations outside, by the sea, or in your garden? Maybe you want to dance it out or chant in true devotion, sing really loudly, or perhaps you just find yourself wanting to pull some serious open and expressive moves. It's good to know that our devotional practice can, and will, change from phase to phase, so be open to it.

Authentic

If there's something you've been feeling vulnerable about— maybe an idea you've wanted to pitch as a feature to a magazine, a class you've wanted to start, or asking the accountant on your team out on a date—now is a really good time to put it out in the world. Why? Because in this phase, you're going to be a lot like the Fool in the tarot card deck, a fearless adventure-ess who is willing to take far more risks than you would be at any other time in your menstrual cycle. In this phase you're likely to be much less sensitive than you'll be in your premenstrual phase, for example, so if you want work critiqued by others or if you want to start something new, criticism won't affect you AS much. I'm not saying you'll be bulletproof—you're human, right? But you *will* be able to take it on the chin *way* more than you would be in the second half of your cycle when sensitivity kicks in.

This is the time to send your book to an editor, to put forward a new idea or concept at work, or to try a new routine with your family. (The last one especially, because you have a LOT more patience in this phase than you'll have at any other time in your cycle too; so if potty training is on the cards, start them on it during YOUR pre-ovulation phase!)

Sensual

Ahh, so in the same way, as we feel a little more flirty in the season of spring, the same is true when we enter our pre-ovulation phase. We naturally feel a little frisky, a little *ooh la la*. So, if you're not in a relationship, this is a great time to initiate one, or at least go on lots of dates, to play, and, most importantly, to have fun.

If you're in a relationship then this is a time to get exploratory in the bedroom–*oh yeah* (I totally said that in a Barry White voice). Don't wait for your partner to suggest sex; your hormones are rising in this phase and you'll be feeling supercharged, so not only will you be more likely to want to initiate sex and play, you'll have the courage, and the ovaries, to be crystal clear about what it is you want too. And I'd REALLY recommend that you *do* get clear about what it is you want. If your needs aren't being met, ask yourself why. If you don't know, get curious about self-pleasure. Take your partner(s) by the hand, literally, and let them know how much you'd *really* love it if they did it like this. Or like that. Keep it playful, ask them to show you what they want, too, because sex study and exploration are HOT.

NOTE

Any problems or issues you overlook in this phase may reappear during the second half of your cycle. This is the way of our cyclic nature and applies to all areas of your life. Your tolerance, patience, and ability to let things slide are higher in this phase of your cycle than any other thanks to rising estrogen levels. BUT if you have a tendency to sweep issues and situations under the carpet at this time (which many of us do) be aware, they may (ha, who am I trying to kid?), they almost always WILL come back to be seen and experienced in the second half of the cycle.

Sensational

Rising testosterone in this phase will not only heighten your sense of adventure but also have you seeking out new possibilities too—people, house decoration, different foods and flavors, different looks, new lipsticks. In this phase, you'll want to mix it up and try new things, but you'll also be more inclined than at any other time in your cycle to spend money on ridiculously not-sensible-in-any-way-shape-or-form shoes. (Or whatever your idea of a totally frivolous but really bloody beautiful purchase might be. Honestly? If you like to shop, when you start charting and seeing when most of your purchases are made, the chances are it will be during pre-ovulation.)

Just as the Moon is growing bigger, so are your energy levels, and your body will be crying out for nourishment. Now a diet, a juice cleanse, or a fast of any kind is not cool at this point in your cycle (I'd argue it's not cool at ANY part of your cycle, but that's a whole other book and totally my personal opinion), as you're going to need delicious nourishment in order to sustain and maximize the infinite energy you experience during this pre-ovulation phase.

When it comes to exercise, I am a belly dancing, yoga, and body movement-loving woman who likes to walk fast and who sporadically breaks into a run. But, because your energy is externally directed at this time of your cycle, pre-ovulation is the time to do any kind of cardio—hiking, running, all that good stuff that gets the blood and endorphins pumping. It's going to feel REALLY good in this part of your flow.

NOTE

What's really great about working with your cycle is that you'll realize that there's not a one-size-fits-all way of working, eating, exercising, having sex, making money, talking to your partner. At each phase in the cycle, our needs are different, so if you've ever taken up running and were like "Hell no, this isn't for me," it might be worth trying it again during your pre-ovulation phase when your energy levels are high and you feel able to take on new challenges. In your premenstrual phase, your body will feel a lot less love for pounding the pavement for 5 miles and will be craving a softer, more feminine movement—did I mention how much I LOVE how this all works?!

You

So, how will you feel at pre-ovulation?

This is potentially where the majority of us feel most comfortable and at ease. This is the phase where you're able to engage fully in tasks, and you have the capacity to deal with the world and be fully in the world because you're externally focused. Beware though, while you may be excited to get started, in those first few days after you've stopped bleeding you may not have the energy levels to sustain how much you actually want to do. You might wake up with a whole ton of energy in the morning, but your body might be crying out for a nap by mid-afternoon. As you move further away from your bleed, in the same way that the Moon is getting bigger as it waxes, your energy levels WILL rise too, and you'll start to find that you *can* work a 13-hour day and *still* feel good-to-go. Don't fret, this isn't a bad thing: this is one of the many, many benefits of working *with* your cycle because when you're in this pre-ovulation phase, you really are able to be in action without it putting stress on your body.

If you run your own business, this is the time to schedule in your marketing and PR, to initiate new collaborations with potential business partners, to write blog posts that can then be scheduled throughout the month because you'll have the energy to do these things with total ease. As we unlock each phase of the cycle, you'll see that during menstruation we have the potential to download a lot of ideas and possibilities, so that when we come into pre-ovulation fresh and renewed, and our energy levels are rising, we have an amazing opportunity to actually start to implement these ideas. We can sow seeds, seek opportunities, and are much more open to the potential of what's possible. This is when we can feel most fearless. So, if you want to put your prices up or ask for a pay rise at work, now is the time to do it because you'll have a sense of *Yes, I'm worth it.* Pre-ovulation is great for giving us a much-needed bravado boost, a you-can-do-it ass-kick. I always try when possible to schedule any public speaking engagements for my pre-ovulation phase, because my energy levels are high and I'm more externally directed; it's easier for me to be around people (as an extroverted introvert who would ALWAYS choose reading a book under a leopard-print blanket over going out partying, I take ALL the help I can get when needing to speak or be out in public!), to engage with others, and to be switched on and able to respond well.

Anything you put your mind to has the potential to get done during pre-ovulation. I used to live my life by an arm-length-long to-do list, and I would get so despondent that stuff just never got done—but now I get it. It was because I wasn't scheduling in sync with my cycle. Pre-ovulation is my implementation phase, a time when things that need to get done *actually* get done.

In pre-ovulation, things take half the time to do. I get far more done and have so much more energy to do them, which makes scheduling my life so much easier.

I personally don't choose to socialize much during this phase because of how much I *can* get done, but when I do, I'm much more open to meeting new people, and nothing really feels like too much of a struggle. There's a chance you won't feel so tired during pre-ovulation either, although I'm sure if you're a mumma you may want to pull me up on that. Even with a baby, I promise you that you *will* get more done during your pre-ovulation phase than you would in any other phase of the menstrual cycle. In fact, I've been known to pull a few all-nighters in this phase—usually when I'm on a deadline—and it didn't feel like too much trouble (she says, crossing all her fingers and toes and praying that after staying up late on edits of this book during pre-ovulation, it doesn't come and bite her on the ass later in the cycle!)

How to get the most out of pre-ovulation

Think before you say yes: As your energy levels build, in the same way as the Waxing Moon grows, your confidence grows, and you'll be much more likely to say "Yes" to life, which is why I need to give you a heads up. While your energy is definitely growing and you may have fearless confidence about you, there will still be some vulnerability, maybe a little naivety too. It's really important to pace yourself and not rush head-first into saying "Yes" and committing to a project/partner too early. I've made this mistake so many times: I'm in my pre-ovulation phase, I'm full of energy, and someone suggests, "Let's do this epic, mammoth, larger-than-large writing project together, and let's only give ourselves a month to do it in, yeah?" And I respond with "Sure, sounds amazing, let's do it," only to find that for the first two weeks, I'm full-on committed and totally into it, but by the third week of my cycle, my energy levels are lower, I'm unable to sustain the same pace and interest, and I want to hide under the duvet from the impending deadline.

I now know how to fully maximize my energy levels during my menstrual cycle (and by the end of this book, you will too) so that now I'll take a little longer to make decisions about the projects I say yes to. I consider time frames, I consider where I'm at in my cycle and when I'm at my most productive and creative, and then I plan my work accordingly. Do I still say "Yes," when I should have said "No?" ALL THE BLOODY TIME. I'm forever in the process of learning, and my cycle is most definitely the most effective teacher/guru I've ever had.

Set intentions: This is a great opportunity to use the clear-headed-ness and clarity that comes in pre-ovulation to set intentions in regard to where you're going and what you're doing this particular cycle. During the menstruation phase, I invite you to look for and be open to receiving signs and messages, but NOT to act on them. In pre-ovulation, THIS is the phase where you can begin to explore them, activate them, and turn them from an idea into action.

Enjoy being a beginner: Enjoy the I'm-just-starting-out energy that occurs in this phase—it allows for a little naivety, incompetence, and mess-ups as you explore and experiment with new possibilities. Basically, this is a really good time to try shit out and fail. The idea of failure can be so all-consuming for so many of us that we never dare to take a risk or to try something different, but this is THE best time: it's the IDEAL opportunity to risk something new or try something different, to dare to have a big dream and allow yourself to face plant. Often. Your resilience levels are super high in pre-ovulation, so while there may be a minor dent to the ego, it will be much easier to shrug off than if you were experiencing it in the second half of your cycle—so go for it, whatever it is, I dare you!

Warning!

There are some pre-ovulation shadows that you might want to look out for each month.

There's a chance that because you're so outwardly focused, you'll forget to check-in with yourself. And there's an inclination to spend the entire first half of your cycle–pre-ovulation and ovulation–running on the masculine "do-er" energy, and forgetting the medicine of the feminine energy you've previously been immersed in during the second half of your last cycle.

There's a chance you'll do too much, just because you can. As I said previously, I'm able to pull a few all-nighters during pre-ovulation, especially when I'm on a deadline, but doing this relentlessly throughout the first half of your cycle could, and most probably will, expose you to painful PMT symptoms in your premenstrual phase. I do this sporadically on the complete understanding that as I move into my premenstrual phase, I need to slow down completely, turn off social media, and allow myself to do less work. (Do I always get this right? No. When I do, does it work? Absolutely.)

NOTE

I totally appreciate I'm not a mumma, and as I work for myself, this "cyclic living" may seem a LOT easier for me to do. But honestly? Over time, it can save time (and provide you with serious support and self-care strategies for life), so let this be an invitation to explore what's possible for you. I have a LOT of mumma clients, and they're working on this pre-ovulation energy ALL cycle. Unfortunately, this creates stress, burnout, and adrenal fatigue, and a lot of them have gnarly menstrual-related dis-ease because of it.

What I've found in working with these incredible mummas, is that when we find a way to work in sync with each phase of the cycle, these mummas get MORE done—they're more effective and yet they're resting and slowing down. When we work with our cycle and not against her, we actually become MORE productive while nourishing, tending, and sourcing ourselves in the process.

There will also be the possibility that you'll have *too* many ideas and may not be able to give enough care and attention to one of them so that it can meet its fullest form. You know when you have a notebook full of ideas, and you try to implement them all, because they're all such good ideas, yet in trying to do them all, you do none of them very well? Yeah, that.

Or, you might be so overenthusiastic and in love with one idea that you share it too soon. For example, one of my clients told her parents, her friends, and her partner that she was planning a brand-new business venture. It was a little bit "out there" compared to what she'd previously been doing, and it was still very much in need of some refinement before launching (refinement is one of our SHE powers that we have access to in our premenstrual phase), but she was just so keen and full of pre-ovulation enthusiasm that she wanted to share it with everybody. Unfortunately, because she hadn't developed the idea fully and didn't have all the answers to the questions they asked, they laughed at her and made some pretty hurtful comments about it, which then led her to want to kill off the idea before it had even started. Luckily, we were able to see how this was impacted by this phase of her cycle, and she allowed this idea to fully unfold and develop over the entire course of one menstrual cycle, using all the tools and powers of each phase, like organization, discernment, and editing. She was then able to view it from each phase, before fully committing to making it happen in a fully-formed, feminine way.

And there's a chance that if you don't use your superpower of getting shit done in this phase, that you simply live in potential, thinking about all the things you *might* do, without actually realizing anything.

 *Rituals*

Yoni steams

Yoni steams are an ancient herbal healing practice to cleanse and nourish the womb. Yep, totally aware of how that sounds. When I told herbalist, Em Tivey about the dark, clotty blood I was experiencing, she recommended I do a yoni steam in my pre-ovulation phase. I admit, back then I was skeptical. But Em is a healer, and she's my go-to-girl for *all* herbal remedies, so I trusted her, tried it, and *loved* it. It's now something I do every cycle. My bleed is lighter, I'm less bloated, and I have no clots. WIN.

What is a yoni steam?

Em says:

Use a blend of organic medicinal plants to make a herbal infusion, and place it in steaming hot water, making sure the steam is at a healthy, won't-burn-your-vulva-lips distance from your yoni. The steam rises and opens and lubricates the yoni with gentle healing, cleansing, and stimulating properties that support a healthy pelvic space and reproductive organs. Vaginal tissue is one of the most absorbent of the entire female body, so the steam opens up the pores of the tissues and increases blood flow to the labia, and vaginal canal, relaxing the muscular and deep fascia layers of the pelvis and womb. It warms, soothes, nourishes, and cleanses the uterus by assisting in the release of both incomplete flushed debris and impacted and old endometrial lining. It also encourages the womb into her natural, open, and upright position.

What would you use it for?

Yoni steams are ideal if you experience endometriosis, very dark blood, or brown fluid at the beginning or end of menstruation. They're also great for irregular periods, painful periods, fertility issues, cervical stenosis, vaginal muscle tension/tightness, ovarian cysts, uterine fibroids, prolapsed uterus, pelvic trauma, or vaginal dryness, or when you're postpartum, post-miscarriage, or post-abortion.

What's in it?

You can choose from a variety of herb combinations, but my favorite is:

Yarrow herb—uterine tonic, pelvic circulatory system stimulant, antispasmodic, blood/liver purifying. Aura protective and highly healing energy.

Marigold petals—uterine and hormonal tonic, healing tissue, softening scar tissue and adhesions, gives a gentle warmth.

Rose petals—most prized of the womb medicines, soothing, softening, tonic, relaxant, cooling, balancing. Aromatic, sensual, loving, and nourishing. Total womb loving!

Raspberry leaf—uterine tonic, nutrient-dense, cooling, antispasmodic, relaxant, nurturing, and protective.

Sweet violet leaf—nourishing tonic to the reproductive system. Sacred symbol of abundance and fertility, and like the womb, a holder of the mysteries of life, death, and rebirth.

You can create your own blend or buy one ready-made from the ever-growing list of suppliers online. I often get asked how to do the steam itself, but there are lots of different methods. Mine? The hot Viking made me a short wooden stool with a yoni hole in it. Classy, huh? I put my herbs in hot water and place the bowl of infused herbal steam beneath the hole. I wear a long skirt, position myself comfortably over the hole without knickers, and let the skirt fall around me to keep the steam in. I put on a facemask too—this is pre-ovulation, the perfect phase for multitasking—and sit for about 15 minutes. However, there are now salons and spas in major cities that offer yoni steams as a service, with private booths and headphones playing soothing sounds if you're looking for a more luxe experience.

Yoga and somatic movement

I have been a yoga/somatic movement teacher and therapist for nearly 10 years now. When I first started, my ass and my belly were very much IN. THE. WAY. My teacher was a dude, and he was awesome. He helped me to find love for the practice of yoga despite my curves. Then I discovered Anna Guest-Jelley, Creatrix of Curvy Yoga, who turned what I knew on its substantially chunky ass and helped me to find a deep love for my curves through the practice of yoga.

Then I met **Uma Dinsmore-Tuli**.

This woman is a powerhouse of inspiration and knowledge, and it's the essence of her work—feminine, divine, and deeply powerful—that now fuels my passion for sharing my own yogic and somatic experience, IN-YOUR-BODY-MENT: a fierce and feminine movement practice that, as the name suggests, invites you to be IN. YOUR. BODY.

I've invited Uma to share why she thinks it's so important to teach yoga with respect and honor for women's health and women's cycles. Uma is a yoga therapist, mother, and feminist writer who's been practicing yoga since 1969. Her book *Yoni Shakti* is my absolute go-to book on yoga for women.

Yoga Was NOT Designed with Women in Mind

So this is how it is: most of the yoga taught today was designed in medieval India, by men, for men. Those guys didn't have menstrual cycles, or vulvas, or breasts, and not much in the way of backsides either, so all these aspects of us are often just inconveniences in most modern hatha yoga (yoga that works through the body to alter the mind). So when we follow yoga instructions intended for male bodies, we give away our power, because those instructions can fundamentally disrespect female bodies.

The disrespect is not simply because the poses we're being asked to do disregard secondary sexual characteristics like buttocks and breasts. A deeper disempowerment occurs when we practice yoga according to instructions that disrespect our body's menstrual cycle by following yoga programs that have no direct relation to our daily experience of our cycle.

It's like this: hatha yoga manuals were composed by men and for men in India between the 11th and 15th centuries. Women appear in these texts, usually only as polluting forces to be avoided. Occasionally a yogi is required to find himself a woman to practice sexual yoga with, involving genital muscular contraction… but beyond that particular technical requirement for a vagina as a receptacle, women's bodies are simply absent.

Think about it for just a minute… would we really expect texts written anywhere in the world in 1450 to be anything other than totally clueless about women and our power? Just because yoga originated in India, and just because medieval Indian yogis were spiritually motivated, doesn't mean that they had any real clue about women's bodies, or about what is a suitable yoga practice for us now.

So although humans of any gender can do the medieval yoga practices, they are not intended for female bodies, and there are certainly no references to menstrual cycles.

I know from experience that honoring and respecting our cycles is a vital means to accessing our own inner guidance, and so it seems to me that most male-oriented approaches to yoga simply encourage women to ignore the inner wisdom that arises when we pay attention to our cycles.

Most traditional yoga schools absolutely discourage asking questions that might reveal a teacher's disrespectful ignorance of women's cycles. These yoga lineages can disempower individuals by valuing the teachings of the "lineage" way higher than any insights from self-discovery. In traditional yoga forms, we are not invited to honor the deep blood wisdom of our own bodies, for fear this might lead us to ask difficult questions of those who would like to be in charge of the wisdom for us...

Such disrespect for women's bodies is surprising, given that yoga is rooted in a philosophy that reveres women and girls. In the Tantric prehistory of yoga, women were often the most powerful teachers, and our sexual energy was honored as a potent creative force. Tantra's respect for female bodies is evident in fabulously body-positive Sanskrit terminology. For example, menstrual blood holds ritual significance, and yoni not only means vulva and vagina, but also "source of all power." A Great Wisdom Goddess is assigned to every part of the vulva, including labia and pubic hair, and menstrual flow is called yoni pushpam, literally "flower of the yoni."

Practices from these pro-feminine roots of yoga, before it got organized into patriarchal lineages, can support us by bringing us back home to our "source power." When we encounter this power, we not only heighten creative and sexual energy, but we access sustainable vitality for everyday life. We can celebrate each stage of life: menarche, menstruation, and menopause as an initiation into deep power. This kind of yoga is a far cry from squeezing yourself into some uncomfortable pose because the teacher says that's the next one in the sequence.

What might that actually look and feel like for the woman who is brand new to charting and understanding life through the blood-red, menstrual cycle lens?

Practical guidance for a vitalizing women's yoga practice

So, what do we do in a yoga class when the teacher instructs us to do something that feels totally inappropriate? How can we practice yoga so that we honor the wisdom of our own bodies, and don't simply follow instructions that disrespect the shapes and rhythms of our physicality? The answer is simple: reconnect to the embodied wisdom that leads us home to our own source power or Yoni Shakti.

There are three steps to access Yoni Shakti: the first is an "internal weather report," an inquiry about yourself and your cycle. The second is to ensure the stream of energy between your capacity to love and your source vitality is flowing well. And the third is to follow your own intuitive guidance about what practice is suited to *you today: the voice of the inner teacher arises when you honor where you are right now and are fully powered by the juicy flow of energy from your heart to your pelvis.*

"Internal weather report": Checking on your five bodies

Don't even move an inch until you have done a comprehensive internal weather report on every body you have. Every body? Yes, indeed, in yoga philosophy, all human beings have not one, but five bodies:

1. **The physical body**–*flesh and blood, composed of joints, bones, muscles, organs. Ask: What hurts? Am I feeling stiff or heavy? Where am I in my menstrual cycle or menopausal journey today? What's going down with my digestive rhythm?*

2. **The energy body**–*permeates and surrounds the physical body, determining levels of vitality. Ask: Am I tired or buzzing? Do I need rest or activity?*

3. **The mental and emotional body**–*creates opinions and prompts reactions. Ask: Am I feeling cheeky, vulnerable, or blah today? Am I likely to giggle or weep?*

4. **The intuitive wisdom body**–*the way we really "know" something when we can't explain it, a special kind of wisdom coming straight from the source–insightful and true. So ask yourself: Am I close to my own truth? Ask: Am I doing what instinctively, intuitively feels right to me right now?*

5. **The "bliss" body**–*our link to the causal state of pure being, and there's nothing to be asked of it.*

Enjoy! But rather than becoming fixated on a particular routine, get radical, get free, and get juicy. When you follow your intuitive guidance that honors where you are in your menstrual cycle today, you can trust that whatever yoga and/or movement practice you choose will be the exact and perfect response to where you are right now.

 *In-your-body-ment*

IN-YOUR-BODY-MENT combines yoga, sound, breathwork, and somatic movement and works in collaboration with the wisdom of the cycles—body, seasons, nature, and the cosmos—to create a fierce and feminine practice that will have you caring very little about the size of your ass and very much about how good it feels to move your body in a way that feels truly nourishing. It reconnects you to your innate body wisdom, soothes your nervous system, cultivates sensorial awareness and pleasure, and provides a safe space for you to explore your inner landscape—emotions, feels, sensations—by moving your body in ways that feel real, supportive, nourishing, and, most importantly, really bloody good!

I've created pose modifications that take into account that we have hips, asses, boobs, and tums, and while I don't encourage you to put your legs behind your ears, you can build strength and flexibility through movement that works with your body's needs and rhythms. Every inperson session is tailored to meet the needs and abilities of the women attending, paying particular attention to where each woman finds herself in that session.

Now pre-ovulation is a time to focus on growth, which is why Ustrasana, or Camel Pose, is a delicious expansive movement to add to your practice.

NOTE

In the practice of IN-YOUR-BODY-MENT, how it looks is irrelevant. If you need to modify anything to suit your body shape, do it. The important part of the practice is to find nourishment in every movement. If it doesn't feel good, keep changing it up until it does. This is your practice, so make it work for you.

CAMEL POSE—USTRASANA

This is a delicious front body, thigh, and abdominal stretch, while at the same time creating a juicy squeeze in your lower back. It opens up your pelvic bowl and gives your entire endocrine system a good flush in prep for ovulation.

Here's how to do it:

1. *Kneel on the floor with your knees hip-width apart (if this is tricky, fold up a towel or place a cushion under each knee) and thighs vertical to the floor. Rotate your thighs inward slightly, stay firm, but don't tense your bum cheeks. Keep your outer hips as soft as possible. Press your shins and the tops of your feet firmly into the floor, so you feel secure.*

2. Rest your hands on the back of your pelvis, palms on the tops of your bum cheeks, fingers pointing down. Inhale and lift your heart by pressing your shoulder blades against your back ribs.

3. Keep your head up and your hands on the pelvis. If you're brand new to yoga, don't drop straight back into this pose, instead, inhale, and with a slight twist, place the left hand on the ball of your left foot, allowing the right arm to stretch up and back and your chest to open wide and strong. Exhale and bring yourself back to the center. Now repeat on the right side. If you're not able to touch your feet without compressing your lower back, turn your toes under to press on the floor and elevate your heels.

4. To move fully, keep your belly soft and lift the front of the pelvis up toward the ribs. Then lift the lower back ribs away from the pelvis to keep the lower spine as long as possible as you press your palms firmly against your heels, keep your neck in a relatively neutral position, neither flexed nor extended, or drop your head back.

5. Stay in this pose from anywhere between 30 seconds to a minute. To exit, take one hand off your heel, swing your arm up and over your head and place your hand on the pelvis, then repeat on the opposite side. Inhale, and lift the head and torso up by pushing the hip points down, toward the floor. Lead with your heart to come up and drop forward into Child's Pose—stay on your knees, bend forward from the hip and place forehead on the floor (or if it's not comfy, onto a cushion) for a few breaths.

 # Herbal healing

After our monthly bleed, now is the time to support the body's process of cleansing. And because of the endometrial blood loss with all of its life-sustaining goodness and stem cells, it's important to replace those minerals and nutrients. The herbs below are the cornerstone for feminine health. Forget diamonds, these green treasures are a girl's best friend!

Nettle (Urtica dioica)

Highly nutritious, dense with vitamins and minerals including chlorophyll, iron, vitamin C, vitamin K, folic acid, protein, serotonin and acetylcholine (both feel-good endorphins), potassium, silica, and calcium. Nettle balances blood sugar levels and promotes convalescence, relieves anemia and fluid retention, reduces heavy menstrual blood flow, and brings on delayed or absent menses. It's also an excellent menopause restorative and blood cleanser, and it increases energy, and increases or decreases breast milk flow. Nettle can be used as a lymphatic cleanser, and a kidney and adrenal tonic for whole-body strengthening during conception, pregnancy, postpartum, menstruation, and menopause. This is a cornerstone for herbal treatment in women's wellbeing, I bet you'll look at this "weed" differently now!

Everyday medicine

Taken as nettle leaf. You can take it daily as a tea (fresh or dried), use handful of fresh leaves in smoothies (pick only young leaf tips without seeds (ideally March to May in the northern hemisphere and September to November in the southern hemisphere—wash well before use), steam as spinach, make soup, or add to savory pancakes.

Shatavari (Asparagus racemosus)

The most important Ayurvedic tonic for women's wellbeing, Shatavari nourishes and cleanses the blood and the female reproductive organs, enhancing female fertility. It also balances hormones and increases natural estrogen levels in menopause, hysterectomies, or oophorectomies (surgical removal of the ovaries), regulates monthly cycles and ovulation, soothes PMS cramps, bloating, and irritability. High in folic acid, Shatavari also helps to prevent anemia, alleviates morning sickness, and is effective after childbirth in cleansing and nourishing the uterus. It increases and improves breast milk flow, and reduces menopausal hot flushes and vaginal dryness, and is also supportive during times of anxiety and depression. It's an adaptogen tonic, so calms and nourishes the body and mind, building immunity, and increasing energy and libido. Shatavari has an affinity with the mind, promoting memory and mental clarity; its calming action reduces anxiety and increases resilience to stress. In Ayurvedic medicine, it's said to promote spiritual awareness and compassion, and is rich in zinc, calcium, and B vitamins. Known as "The Queen of Herbs," her name translates as "She who has a thousand husbands,"—which gives a clue as to its power and effectiveness as a tonic herb!

Everyday medicine

Taken as Shatavari root—daily as capsules or dried powder in a smoothie.

 ## *Flower power*

Daffodil

Keywords—Hope, new beginnings, spring, self-esteem, clarity, and communication.

The essence of daffodil is magical. When we feel we need to find that impetus to move forward and find direction, daffodil can help. Daffodils allow us to be clear on our intentions; they help us to enjoy life while shining our light. If you feel doubtful or full of negative self-talk, daffodil will help you to find YOUR voice. If you struggle to project yourself, this flower has a gentle illumination to it, which always inspires and makes us feel fearless and able to communicate with clarity and love. Use daffodil essence during pre-ovulation or to support any kind of new beginning.

CHAPTER 5

Ovulation–Queen of Freakin' Everything

Female archetype: Mumma/Creatrix

Moon phase: Full Moon

Season: Summer

Element: Fire

Tarot card: The Empress

Crystals: Rose quartz–soft and embracing

Amber–supports self-assurance, balances masculine and feminine

Essential oils: Lavender–relaxes, calms, soothes all

Rose–purifies; physical, emotional, sexual

Mantra: "I can do ANYTHING."

Song: *Girl On Fire*–Alicia Keys

Perfume: Yves Saint Laurent–Black Opium because it's pretty damn sexy, which is exactly what you are during the ovulation phase of your cycle–purr!

The science bit

At ovulation, a tiny egg has been released from one of your two ovaries. So if you're a woman who digs on dudes for sex, *this* is the time you're most likely to get pregnant—you've been warned. Wink.

This is the second phase of your cycle and usually begins around day 13 all the way through to day 21, during which time it may feel like you have a neon sign above your head flashing: *I'm hot-to-trot, come and get me!* and that's because, g-friend, you're ovulating.

Since day 1 of your cycle, estrogen and testosterone levels have been continuing to rise and collaborating daily to create good feels, high energy, and an upbeat nature, so that you'll either hook up with a man or overlook the bad habits of the one you already spend time with. It also means you're more likely to keep him around until at least day 13 when the possibility for pregnancy is super-high in women who don't use any form of contraception. If dudes aren't your bag, your biology will still be sending out all the *I want to reproduce* hormones because... well, that's what biology does. Ovulation happens when estrogen and testosterone reach their highest peaks during your cycle. This means that the optimism, confidence, personal power, and the "doing" mentality that have all been growing since day 1 of your cycle are now peaking too.

Queen of freakin' everything

Like the Full Moon when she's ripe and has reached her full potential, during ovulation you become your most full, and your most present. It's a time when you become pregnant with life. Some women literally make babies, while others become much more nurturing, compassionate, and a whole lot more passionate at this time. You may experience a strong need to reach out and work, collaborate, make out with, and generally *be* with people. Even the most introverted among us may find it much easier to be with and around people during ovulation than at any other time in our cycle.

The good news is if baby-making isn't on the cards right now you don't have to become a baby mumma to be a Mumma Creatrix. This phase is when you're at your most creative and most productive. You're also at your most sociable and your most expressive—and *these* combined? Well, they're the key ingredients to manifesting the very best kind of life magic.

Yep, at ovulation, you become a manifesting maven.

If you're fully in sync with your cycle, there will be moments during the ovulation phase when it will feel like you can click your fingers, make your demands to the universe, and be provided for within the hour—like an insta-manifest drive-thru. The magic of the menstrual cycle means that every menstrual month you've the opportunity to reap the rewards of the seeds you planted and the actions you took during pre-ovulation; to create, to manifest, to take the opportunities and potential of the pre-ovulation phase and use them to become the mistress of your destiny.

That's not all. Everything can feel like it has a delicious rosy hue like you're wearing rose-tinted glasses, and you feel like you're in love with it all—people, places, trees, nature. The Mumma archetype may have you feeling like you want to fix everything because a mumma's natural instinct is to focus all her energy on her offspring, whether that's actual children, her pets, her garden, her friends, or her creative projects.

Now, this is gorgeous and delicious energy, but what I want to invite you to explore in your journal is what the term "Mumma" means to you, so that when you come into this ovulation phase, you'll have an idea of what might come up for you. For some women, they may have had overbearing, domineering, and interfering moms, so the ovulation phase can leave them feeling suffocated and like the very last thing they want to do is be around other people. Others may not have mothers or have had a less than positive experience with their mother or of being a mother, and feel like they have to learn from scratch to nurture themselves as well as those around them.

For one of my clients, ovulation was a total battle. She wanted so much to be a mumma that during the phase when she should have been at her most fertile, instead of exuding the Queen of Freakin' Everything energy—her full, illuminated potential—she became stressed and full of angst. She wasn't able to mother and nurture herself through the phase; she showed herself no compassion and instead blamed herself for not being good enough to be a mom. We worked to turn that around by exploring some of the deep beliefs the Mother archetype brought up for her, and by changing those, she was able to find love for this phase of her cycle again. She became less stressed and was able to tap into the powers of compassion and passion. She also went on to have a beautiful baby boy.

This is a very complex subject, and one that I totally understand can't be tackled in a 15-minute journal prompt. But I'm inviting you to begin to explore your personal relationship with the concept of the Mumma archetype, so that as you enter the ovulation phase you'll start to have an awareness and a knowledge of what might show up and how. So get still, set a timer for 15 minutes, and journal straight from the heart. Include your ideas of what a mother "should" be, and what your mumma is/was like? What are you like as a mumma? If you're not one, how do you feel about that?

Ovulation is like summer, the pagan Sabbat of Litha, Midsummer. It's the time of year when the crops are growing and the earth has warmed up. Days are long, and we can be in nature. Like the Empress in the tarot deck, you're the epitome of feminine power during ovulation; fertile, sexual, fecund, and abundant. The Empress is the archetypal Earth Mother, the Feminine Principle. She is Goddess Demeter. She is Freyja, the goddess of fertility who represents the creativity, fertility, art, lust, passion for life, harmony, luxury, beauty, and grace that's all available to you during this phase in your menstrual cycle.

Your ovulation SHE powers

Each phase has a set of superpowers you can access and hone to make life a whole lot sweeter during this part of your cycle.

At ovulation, you've a presence because it's more possible to be present. You can be totally captivating in *any* situation. Think Marilyn Monroe in *The Seven Year Itch*. I know, right? I can guarantee Marilyn was ovulating when she did the scene with THAT dress and the air vent.

You can take on the world, superwoman style. (Lycra—as always when I talk of superheroes, which I do a LOT because I'm a fangirl—is optional.)

People can assume that you're one of those "lucky" people in this phase because, well, things just naturally go your way. This is especially true on social media because the very nature of ovulation means that you're more likely to post and share more during this phase than any other, so those who follow you, receive your best and fullest expression through their screen.

During ovulation you have the potential power to conjure up your dream job, ideal clients, or maybe even a cheeky hot date. In fact, whatever you really desire, there's less doubt that you can make it happen during ovulation than any other phase of your cycle.

How to activate your ovulation SHE powers

You've access to these particular SHE powers during the ovulation phase of your menstrual cycle EVERY. SINGLE. MONTH.

The best way I've found to fully activate them and use them to their full potential is by aligning your SASSY—Spiritual, Authentic, Sensual, Sensational YOU—but remember, it's up to *you* how you choose to use and interpret the information in each phase. This is just simply a guide to help you on your way, OK?

Spiritual

During ovulation, you can feel spiritually out of sorts. Why? Because your focus is still very much outward-facing, as it was during pre-ovulation, so the idea of meditating or simply being still can feel uncomfy and like you'd rather be doing something/ anything else rather than this. The good news is SHE/universe/ source is present everywhere. Just because you're focused externally doesn't mean you can't have a spiritual practice. You simply find different ways of expressing yourself spiritually (which is why if you've studied yoga and practiced *Sadhana* for any length of time, you'll know that this is a very masculine concept; feminine energy is just not made to be contained or tamed in this way). If your normal spiritual practice is to keep still and have your eyes closed, there's a chance you'll want to resist it at this time. Instead of ditching the practice altogether and berating yourself for not being "spiritual enough," maybe now is the time to open your eyes, to sing or dance or shake. Instead of silent meditations, turn song and dance into a devotional prayer.

This isn't inhibiting your spiritual practice. Lots of my gorgeous clients worry that they're not doing spirituality the "right way" if it's not super-deep and profound.

Bullshit.

SHE, the Divine, whichever deity you choose to give your love and devotion to, doesn't care how you express yourself, just as long as you *do*.

You may question why you feel disconnected at ovulation, or why your spirit guides are harder to contact at this time. Or you might just be thinking, *Let's sack off the spiritual practice, I'm really hot for sex right now!* Make THAT your spiritual practice.

These highly sexual feelings are TOTALLY normal during ovulation, so find a way to connect the sex and the spirituality for you—they're not separate. In fact, a sexual practice can be a really powerful spiritual practice, either solo or with a partner(s) that totally gets it.

I. HIGHLY. RECOMMEND.

Authentic

You're not so easily swayed during your ovulation phase, and you'll feel like you're able to do what you want to do with a lot more guts and conviction. In fact, you'll feel pretty invincible, like you're Super Woman AND Wonder Woman combined. Like you can do anything, and that's because, well...you potentially can. But let me be clear, it doesn't mean that you *should*. In fact, I invite you, in order to fully benefit from your Mumma/Creatrix energies, not to force things to happen but to ALLOW things to happen.

Back in your pre-ovulation phase, you planted the seeds of possibilities, and in ovulation, you're now able to let them manifest *through* you.

So, take your hands off the wheel and let life, your desires, your creative energies move/dance/express/come in to being through you. Be a Co-creatrix of life by being the instrument through which it can flow. Lots of people think manifesting means DOING, and yes, there definitely has to be action, but let it be in communion with SHE/universe/source, with your life force. Trust that what you planted, the opportunities you took, and the possibilities you followed up on, can be manifested now during ovulation.

Sometimes it seems that in order to really *feel* like we're doing something, we have to start something new. We suffer from FOMO (Fear of Missing Out)—we might have done a whole bunch of amazing things, but we'll make like a magpie and see something new, something bright and shiny, or we'll see someone doing something completely different on social media, and we'll think, *Ohhh, I should be doing that too.*

STOP!

You really *don't* need to start doing that too.

Follow through on the actions you took in pre-ovulation; you're growing and building something. Stay with it, and use the Mumma/Creatrix energy to nurture and support it, because this is a precious part of the process.

You'll also feel a lot of acceptance toward other people in this phase of your cycle. At other times of the month, you may feel annoyed or agitated, and that you just can't deal with people or certain situations. During ovulation you'll be a lot more accepting of people's issues, and of their behavior. Things that would normally irritate you, you'll find yourself being totally fine with—it's actually a pretty awesome feeling. Which makes it a great time for healing conflict. If you're arguing with a partner or a friend or someone at work, now is the time to deal with those problems and have *those* conversations. You're more outwardly focused, so have that dialogue now. It won't feel as personal as it might if you were to do it next week, for example, when you head into your premenstrual phase, when your tongue will be sharper and your truth uncensored and untamed. Be cautious, though, because your heart is pumping out BIG love right now—it's tender, so pay attention to it. You'll be longing for connection. Not just physical and sexual, but also in friendship and love.

Sensual

Your libido, your sexual appetite, is potentially going to be off the charts right now, and there's a GOOD chance that during ovulation, you'll be feeling like *I want to have sex.*

I say honor that, g-friend.

If you're in a relationship or in a partnership, let your partner(s) know what you're feeling. Don't be shy, don't be inhibited. Let them know what you desire most. You're in your fullness at ovulation; you're emitting some heavy-duty arousal-inducing pheromones, and with estrogen and testosterone peaking, sex will feel GOOOOOOD.

You lubricate easily, you're quick to be aroused, and climax is easy to reach. And once reached, orgasm can be an intense and all-over body experience. If you're currently solo, honor this potent sexual energy by cultivating a delicious self-love masturbation practice—yep, I totally mean touch yourself. In fact, if it feels good (and I'm aware that for some women, orgasm, for many different reasons, isn't comfy) but if you're open to exploring, I invite you to experience orgasm pretty consistently throughout ovulation. Fingers, crystal wands, Betty Dodson's barbell, or a multispeed vibrator—I recommend them all.

Give yourself permission to experience sexual pleasure in whatever form feels good to you during this phase of your cycle. It's not the only time you can do this, obviously, it's just that this is the manifest-awesome-sex time. Can I get a high five?

What's that? You've not got a hand spare?

Good work!

I also recommend using the act/art of orgasm to manifest. It's one of the ancient feminine temple arts and I'm ALL. ABOUT. IT.

SELF-PLEASURE PRACTICE

Now, this isn't about me telling you HOW to bring yourself to orgasm, and you definitely don't have to do this to manifest during this phase because your manifesting nature is heightened simply through the act of ovulation. But if you're comfy with self-exploration, I do recommend a li'l self-induced pleasure with the added bonus of boosting your power. Here's a practice to get you going.

1. *Tap your heart three times.*

2. *Touch yourself in a way that feels goooooooood.*

3. *Close your eyes, bring your attention to what it is you're calling in, and let visions and sensations make themselves known to you as you orgasm.*

4. *Enjoy yourself, play with yourself, and enjoy and play with your partner(s) too. Don't hold back.*

If, after the blatant permission slip to get it on, you're thinking: *I'm still horny, what's wrong with me?* There's NOTHING wrong with you, you're totally in flow with your cyclic powers—use them for good; manifest orgasms, multiple orgasms, all-day orgasms, ALL the orgasms. Orgasms are good. FACT.

Yes, yes, yes, I'm also aware that sex isn't ALL about the orgasm either. My point is, this a time to make arousal and pleasure—YOUR arousal and pleasure—YOUR priority.

Sensational

So how are you going to feel during ovulation?

Estrogen and testosterone are at their peak, and your energy levels are as high as they're going to get in your cycle, so now is the time to do anything that requires stamina, endurance, and drive–think a 10-mile hike or moving house.

Personally, I'd take the all-day, super-luxe sex session over either of the above, that's just me. But know that your pheromones make you a supercharged magnet, so whatever you set your heart and mind to during ovulation, you're able to attract it to you.

I don't know how scientific this is, but in many of the online cycle programs I share with women, I ask them to take selfies in each phase so we can explore our feelings toward ourselves in each phase. Virtually EVERYONE will say that they prefer the image of themselves that they took during ovulation over every other phase. It's definitely true for me that I feel more "attractive" to the outside world during ovulation than at any other time in my cycle, so explore it for yourself. Not as a way to judge yourself, or to feel "good" or "bad" about how you look, but as a way to get curious about your own thoughts, feelings, and responses to your appearance in each phase.

Estrogen and testosterone were in charge from day 1 through to day 13, now it's progesterone's turn. Progesterone is going to dominate from now through to your bleed, and the thing about progesterone? It's a feeder. Yep, you will eat roughly 12 per cent more from day 14 through to day 28 than you did at the beginning of your cycle. Why?

Progesterone creates a sensitivity in your blood sugar levels, which means you'll feel hungry every three to four hours instead of every five to six.

So, pay attention to your tummy. Eat more regularly. Otherwise, you'll find that you might be irritable and get attacks of the hanger (hungry anger)—tell me this isn't just me?—as you move from ovulation into premenstruation.

You

If you let it, there can be a really delicious ease to life at ovulation. If you've set your intentions during the pre-ovulation phase, now will be a time for them to be created, nurtured, and made fertile. Yogis call this your *Sankalpa*. When you set a *Sankalpa*, you send out powerful vibrations into the cosmos to create a shift in your life. Your *Sankalpa* allows you to let go of, and dissolve old, constricting habit patterns of the mind. It allows you to create alignment in your thoughts, words, and actions to manifest your deepest, most real-to-you desires.

As you move from pre-ovulation to the ovulation phase, the vibrations of your *Sankalpa* will grow stronger, and you can begin to experience its power. You'll see it start to blossom like the lotus flower and manifest in your life.

I invite you once again, though, to allow natural growth to happen. Don't push, allow. Use the solar/masculine energy that you've access to during this phase to create. Don't switch to something else shinier, or more appealing. Don't start something brand-new during ovulation either, because if you start all over again here, there's a chance you won't have the energy to sustain it through to completion as you begin to move into the second half of your cycle. Starting over could lead to crash and burn as you move into premenstruation, and nobody wants that, right?

How to get the most out of ovulation

Let it flow: At no other point in your cycle are you *more* in flow and fully in your power, so for goddess' sake, let it be easy. (I seriously think we all need to make this our life mantra.) At ovulation this manifesting energy is potent, so let lady luck be your guide in this phase, because when you do, life can get really, really good!

Be productive: Save anything that you need to do that feels uncomfortable or stretches you—like a new product launch, or the delivery of a presentation or workshop—until ovulation because you'll have lots more energy to create the best possible outcome. Batch cook meals (because I assure you, come premenstruation you'll want to kiss the face off your ovulation-self for thinking of it), host a dinner party, create and share content, and follow up on work leads because, quite frankly, you're hard to resist when you're ovulating.

You're magnetic: Be careful what you wish for in pre-ovulation because there's a REALLY strong possibility it'll come true during ovulation. If you're launching a book, a project, or something that needs you to be available publicly, then make sure, wherever possible, your release date is during your ovulation phase because you're just so much more magnetic and able to show up–physically, verbally, socially, and psychologically.

Warning!

There are some shadows in ovulation that you might want to look out for each month.

We have the ability to be an all-singing, all-dancing superhero during ovulation; the high energy, the attention, and the fire of this phase can be totally intoxicating and lead us to just keep doing, and creating, and doing and creating 'til we burn out and have no energy left. So keep checking in with your sweet self during this phase of your cycle. Slow down and take some deep breaths, be your own mumma, and nurture yourself before nurturing projects.

You may be SO driven to get shit done that you neglect yourself, a friend, and/or a partner in the process. This is NOT cool. You might be so into making it happen that you forget why what you're doing really matters. You may even ask yourself: *Why am I doing this again?* So be cautious.

If you're a natural introvert, things can seem a little *too* bright during ovulation, like you need to break out the Ray-Bans. This is totally understandable; the heat to perform and be seen in this phase can feel harsh and relentless if you're not prepared for it, so be sure to get really clear as to what's a "yes" for you and what's a "hell-to-the-freakin' no." Setting boundaries for yourself and for those around you are really important in this *I can do anything* phase of your cycle.

 # Rituals

Self-love

For a woman to truly be able to harness her powers at ovulation, she needs to have big love for her belly. Whether it's a belly that pops out from under your vest in Downward Dog or a belly that's not held in for photos–a belly that's allowed to breathe, be seen, have fire, and express itself fully really is the best kind of belly.

FIND LOVE FOR YOUR BELLY

Look down at your belly now. Take a deep breath in and then out. Allow her to simply relax in her natural state. This might be a little difficult at first, and it may take a few goes, because we've been programmed to keep her hidden and to suck her in, but allow her to be expressive–don't judge her. Place both hands just around your belly button and simply say, "I love you, belly." It may start as a whisper–I know mine did when I first started to do this–but she's home to your fire, your wisdom, your instinct, and it's time to show her some serious love. You may want to make judgments about how she looks or feels, but right now, all you have to do is to tell her you love her.

Do this EVERY DAY.

Make notes in your journal on what feelings come up for you when you do this. The first time I did it, I sobbed. I'd hated on my belly for

SO long–the abundant rolls of lily-white flesh–that I didn't know how to even begin loving it. It may feel silly, it may feel simple. Whatever resistance comes up around doing it, do it anyway.

Become a manifesting maven

So, I've spoken about how during ovulation, you quite literally become a manifesting maven. Working with my cycle and the cycles of the Moon, I have manifested book deals, vacations, and, most importantly, the hot Viking during a Full Moon and ovulation.

How?

In its simplest form, get really clear about what you want to manifest, and then you make a list.

Yep, you write a list.

Write down what you want. The best way to do this is to write in the present tense and to start with the sentence: *I am so grateful for...* After that, you can fill in the blank. BE SPECIFIC.

Now, assess your actions and behaviors, because while *The Secret* may have claimed to have *all* the answers to manifesting, it didn't mention the most important part: that you actually need to take action. If your actions and behaviors are *not* in alignment with what it is you want to manifest, SHE/universe/source can't be sure that it's what you REALLY want. For example, if you want a book deal, but you're not actually sitting down to write, your chances of getting a book deal will be slim to none. If you want to invite love into your life but you're always busy at work, staying late and working long hours at the weekend, then there's no opportunity for love to find you there. So assess your actions. If you want the book deal, show up and write, even if it's only for 20 minutes a day. If you want to invite love in, clear some space in the diary, sign up to a dating app, and ask your friends to keep you in mind when they meet new people they think you might be attracted to.

What we think, eat, drink, say, watch, listen to, read, and focus on ALL affect our powers of manifestation, so be sure that the action you're taking is going to up your good vibrations and make you super-magnetic to ALL of the good stuff.

 In-your-body-ment

This one is called the Hare Pose because supposedly you should resemble a hare when you're in the final position. I'm not entirely convinced about that, but I *do* know that it's a pretty divine stretch. And if you experience pain at ovulation, which a lot of women do, a nice little modification would be to create fists with your hands and place them into your groin as you fold forward to offer a little support.

HARE POSE–SHASHANKASANA

This pose creates a space of rest during what can be a phase of heightened energy. Bringing the body forward in this way gives the reproductive organs a loving squeeze and activates the rest and relax mode of the parasympathetic nervous system. Sigh.

1. *Sit on your knees. Place your hands on your thighs and breathe.*

2. *Inhale, and raise both your hands above your head, palms facing forward. The arms should be in line with your shoulders.*

3. *Exhale, and slowly bend from the hips toward the floor and bring your hands forward until your hands and forehead touch the ground.*

4. *In the final position, the forehead and hands rest on the ground. Rest in this position for as long as you're comfortable, taking slow, rhythmic, and relaxed breaths.*

Herbal healing

At this time of inner summer, our bodies want to be in full fertile swing, on many levels. Physiologically, these herbs support the ovaries and pituitary gland to finely tune the hormonal feedback systems that make our bodies cyclic. And this is important whether you're interested in conception or contraception.

136

Vitex or chasteberry (Vitex agnus-castus)

A Mediterranean shrub with the wonderful ability to stimulate and balance the function of the pituitary gland and support the ovaries in the second half of the cycle. Supportive in PCOS, this herb also regulates cycle length, relieves emotional distress of PMS, regulates irregular and painful menses, helps with heavy bleeding, pain mid-cycle, and fibroids, and re-establishes hormone balance after coming off the pill. Excellent for lactation, menopausal depression, and hormonal imbalances.

Please be aware this herb enhances fertility, so if you don't wish to conceive a child, use contraception.

 *Flower power*

Gerbera

Keywords—Optimistic, expressive, bonding, open, clear expression.

You can't get cheekier than the essence of gerbera. These flowers are wonderful in supporting us to express ourselves. They help us to open up and allow us to be in our power and fullest expression. It encourages sociability, fun, joy, and lightness, and invites in friendships and bonds that will last a lifetime.

CHAPTER 6

Premenstruation– Charmed and Dangerous

Female archetype: Wise, Wild Woman

Moon phase: Waning Moon

Season: Fall

Element: Water

Tarot card: The High Priestess

Crystals: Obsidian—protecting and transforming, brings light to dark areas

Carnelian—for clearing the mind, communication, and expression

Essential oils: Geranium—fluid retention, balances hormones

Neroli—calm, stress-reliever

Frankincense—release and let go of past, evoke spirit

Mantra: "I'm charmed AND dangerous."

Song: *Respect* by Aretha Franklin

Perfume: Chloé Love Story, or Eau de Clarity as I like to call it, perfect for when you need to see the truth or make decisions

The science bit

This is the third phase of your cycle, and if your egg wasn't fertilized during ovulation, from roughly day 23 through to your bleed, you experience a withdrawal of the following three hormones—estrogen, testosterone, and progesterone.

Now, decreasing estrogen can activate anxiety and nervousness and reduce serotonin (the hormone that stabilizes your mood) . When this happens? Noradrenaline is produced.

If something irritates you—like your partner leaving wet towels on the bathroom floor (please say that this isn't just me?!)—it's a sure thing that your partner is about to hear ALL about it.

Meanwhile, decreasing progesterone levels mean that you're probably crying at… well, just about anything. And after the high energy levels, you've been experiencing in the previous two weeks, decreasing testosterone will mean that doubts are setting in. What about? EVERYTHING. Is it any wonder so many women hate this part of their cycle?

The good news is that, despite what many women experience in this phase—bloating, sore breasts, spots, headaches, cramps, mood swings, and the *I'm-not-good-enough* feelings—this phase *does* have superpowers, you *can* find love for it, and for some, including myself, it can end up actually being the phase of your cycle that you look forward to the most.

Honest.

Charmed and dangerous

Up until this point in our cycle, playing by societal rules has definitely had its benefits. Our energy is directed externally: we can meet deadlines, complete tasks, and we can use the fiery solar power to fuel our confidence and self-esteem. Except, as we move into the premenstrual phase of our cycle, the masculine, logical, straight-line thinking and practical traits that served us so well in the previous two weeks, now become a very limited tool kit as we enter the feminine-led phases of our cycle. As you move into this part of your cycle, you may experience you're "doing", I-like-to-achieve self wanting to do all it can to remain in control, but there's a shift from *doing* to *being* in this phase.

Unfortunately, this shift is often ignored by the all-singing, all-dancing, all-doing superwoman that we cultivated during ovulation. She thinks she can, and should, simply keep on pushing through and functioning at the same levels of high energy that she has been for the past two weeks, but hormonally, it's just not possible—at least, not without some serious side effects. These include everything from irritability, frustration, confusion, and sadness through to depression, anxiety, and addiction.

This phase is often overlooked and gets a bad rap, which for me, is particularly interesting as it's represented by the Wise, Wild Woman archetype; the one which modern society ignores in favor of a younger and more vital version of womanhood.

Like the High Priestess in the tarot deck, the Wise, Wild Woman represents wisdom, serenity, knowledge, mystery, and understanding. The High Priestess represents spiritual enlightenment and inner illumination, divine knowledge, and wisdom. She has a deep, intuitive understanding of the Universe and uses this knowledge to teach and share, rather than to try to control and manipulate.

The thing is, this phase can get a little messy, and as modern women, we learn very early on in our lives to keep our "messy" aspects under wraps. Our emotions, should they spill out, get squashed down, and we apologize for them. A LOT. We worry that we're being seen as "too much", while at the same time, we're struggling with feeling that we're not enough. This is indicative of the virtually-never-spoken-about phase of perimenopause, the transitional phase between our bleeding years and menopause. On her blog for MPowder, Vera Martins describes perimenopause as: "a time of heightened hormonal instability. It's like your hormones are on a rollercoaster—think of it as reverse puberty."

Because it's not spoken about, perimenopause has many women feeling like they're going mad and/or crazy, which is also how many women feel each month as they experience their premenstrual phase. We've been tamed, shushed, and censored, and it's never more present than in this part of our menstrual cycle because it's here that our wildness—our truth, our voice, our body, our very essence in all its messy imperfectness—demands to be untamed and uncensored. Now, if that *actually* happened, if we took life by the ovaries and actually spilled out our thoughts and feelings, our truth, direct from our heart and without censorship—a heavily patriarchal structure, like the one we currently live in, would be scared shitless. Because, quite frankly, we'd be dangerous.

Except we wouldn't be, not really, we'd simply be living in our truth. Our fullest expression. What's more dangerous to me is that we've disowned the second half of our cycle. If the first half is a deep inhale, this second half is the exhale; the let go, the opportunity to stop doing, and to really be. To let our body and our heart lead us into a different, deeper way of being. And yet we're totally disconnected from it. Until now, that is.

The Wise, Wild Woman archetype invites us to explore our natural instincts. This instinct is a deep knowing. It's where we should feel most at home, yet when we go against our cyclic nature, which we do if we ignore or disconnect from the premenstrual and menstrual phases of the cycle, we suffer. We become so disconnected from our knowing as women, the knowing that was previously celebrated and revered, that we no longer recognize how to return to it. Yet this is EXACTLY what this premenstrual phase is for. To call us back into our body, into our truth, so we can be present to ourselves, and for ourselves, when we bleed at menstruation.

What's your instant response to the Wise, Wild Woman? How does she manifest in you? Do you suppress her? Does she feel so uncontrollable you have to keep her on a leash? Have you even met her yet? I invite you to get still with yourself, set a timer for 15 minutes, and allow yourself to write from the heart. When I first explored my Wise, Wild woman, I was scared of her; I thought she was "too much", but that's what we've been programmed to believe by society. But slowly, with each cycle, I've started to hang out with her to really explore her rage, her bitchy-ness, her uncensored nature, and she has become my mentor in owning my voice and speaking my truth.

Be warned, not everyone will dig your Wise, Wild Woman, especially if you've never invited her in before. At first, she may feel overwhelming, but take it slow and drink her in like a fine Mary Magdalene-approved red wine. Like fall or around the pagan Sabbat of Samhain, the time of year when the trees start to shed their leaves, and when you feel the need to be inside more, preparing for the winter, premenstruation is an opportunity to look inward and to care less about what's going on in the outside world. Instead, turn your attention to your own inner landscape, to experience your Wise, Wild Woman fully.

Your premenstruation SHE powers

Here's your set of superpowers that you can access and hone to make life a whole lot sweeter during this part of your cycle:

- You can trust your intuition and self-knowing to guide you in making big-ass life decisions.

- Your bullshit detector is set to high.

- Your psychic abilities are heightened—yep, and you don't have to wear a headscarf and big silver hoop earrings to do it either. Unless you want to, obvz. I do.

- You're able to make bitchcraft–the act of speaking your mind and truth–a positive thing. Because speaking your truth doesn't have to rude or offensive.

- You become an editrix. You can spot problems in work, in relationships, and throughout your life, and you know what needs to be done to fix them.

Good news: When you work with your menstrual cycle, and not against it, you get the opportunity in this premenstrual phase to be charmed and potentially a li'l bit dangerous (and when I say dangerous, I'm talking about the truth-telling, Wild, Wise Woman kind) EVERY month.

Yep, for the 8-10 day period of time, each cycle, from day 21 through to the day that you bleed, you become a truth-seeking, cut the bullshit, mistress of bitchcraft–if these powers aren't harnessed, there's a chance they can manifest as pain, anger, or perceived craziness. But if you're able to work with them?

You become a supercharged, power source for truth.

Your intuition and psychic abilities heighten, you become charmed and dangerous in the very best kind of way.

How to activate your premenstruation SHE powers

You have access to these particular SHE powers during the premenstruation phase of your menstrual cycle EVERY. SINGLE. MONTH.

The best way I've found to fully activate them and use them to their full potential is by aligning your SASSY—Spiritual, Authentic, Sensual, Sensational YOU—but remember, it's up to *you* how you choose to use and interpret the information in each phase. This is just simply a guide to help you on your way, OK?

Spiritual

In your premenstruation phase, if you allow it, your connection to spirit is high. In fact, it can be almost overwhelming how deep you can potentially go here, connecting with everyone from ancestors to guardian angels. Use this time to work with your crystals or to buy an oracle deck or to have a tarot/psychic reading—because the closer you get to menstruation, the thinner the veil between you and spirit will become, and you'll have access to massive amounts of insight and wisdom.

Don't be afraid to pull cards for yourself at this time for guidance. You don't need to, but I find oracle cards to be super supportive at giving my heart and my intuition a little cosmic nudge. For my daily readings, I love to do a three-card pull from either the *SASSY SHE* or *SHE Sirens* oracle decks, mainly because I'm a witch, and as De la Soul so correctly surmised, three is indeed the magic number, but also because a three-card spread cuts the crap and gets straight to the heart of the matter. My favorite placement is:

1. Past—What past experience has shaped your current situation?

2. Present—What's going on for you right now?

3. Future—Where is the situation heading?

GO WITH YOUR SHE FLOW CARD SPREAD

When I created The SASSY SHE Oracle Cards, *I also created a spread to guide you through each menstrual cycle called:* Go with your SHE Flow. *You don't need the SASSY SHE cards to do this layout (although they are really fabulous, and if you do want a set, you can buy them direct from www.thesassyshe.com). You can use any tarot or oracle cards you're particularly drawn to.*

1. *On day 1 of your cycle, the first day you bleed, pull five cards.*

2. *These cards represent the four phases of your cycle— menstruation, pre-ovulation, ovulation, and premenstruation. The fifth card will provide insight into the entire menstrual month.*

3. *Each card will let you know what guidance you need at each phase to give you a little helping hand throughout the cycle month.*

Just like those three kickass witches on the TV show of the same name–jeez, I LOVE that show–in your Wise, Wild Woman phase you are charmed. You're able to see what's REALLY going on with both yourself and other people, and you see straight through the bullshit, direct to the heart of the matter. This alone is one of my favorite parts of the cycle because when THIS power is harnessed, no one can lie to you without you knowing–people will think you're reading their mind, but it's because your intuition is simply heightened. It's a great time to journal, to go deep and let your inner knowledge reveal itself. I'm passionate about journaling (I call it heart riffing: letting my heart riff directly onto paper. Yep, I'm old school; I still use a pen and paper) as a spiritual practice because signs and messages will be able to come to you at a crazy speed. This is the perfect part of your cycle to get still, to create a beautiful altar dedicated to yourself, to your cyclic nature, and to meditate with your journal close by.

When I'm in my premenstrual phase, I like to take myself on a journey with the beat of my drum or tambourine. I'm from traveler descent, so playing these instruments is very much part of my personal spiritual practice. I'll set a clear intention in my journal saying something like, "I am journeying to seek support," and then I'll allow myself to be taken with the beat of the drum. After about 20 minutes, I will stop and heart riff my discoveries and what's unfolding–what I see, hear, and feel–into my journal. It's really powerful. I know that some people can't journey while drumming themselves, so if this isn't for you, you can get recordings online.

I also LOVE Yoga Nidra. This is something the Viking teaches in his Radical Rest sessions, and he and I have been huge fans and advocates of this practice for years now, mainly because it really bloody works.

Yoga Nidra is basically, yogic sleep. It's among the deepest possible states of relaxation you can enter while still remaining fully conscious. I know, right? Quite frankly, it's bloody delicious. I've not always found meditation easy, but Yoga Nidra creates space in my mind and body to really deeply relax. If you ever come to a workshop run by either the Viking or me, you'll experience it, and it's part of every IN-YOUR-BODY-MENT session. I've created Yoga Nidra experiences for each phase of the menstrual cycle that can be downloaded and listened to just before you go to bed (for sale at www.thesassyshe.com/shop). Ahhhhh. I'm feeling blissed out, just thinking about it. Sigh.

Authentic

During our premenstrual phase, especially the first few days, we can feel like we're riding a rollercoaster–high energy one minute, followed by "I need a nanna nap, pronto" the next. This can feel both confusing and frustrating, having just come out of the high-vibe energy of ovulation. It can also be confusing for those we live and work with because, in the same way that the Moon wanes in the second half of its cycle, our hormone and energy levels start to drop in premenstruation, and we may no longer have the stamina, nor the inclination, to commit to what we said we were going to do back in the first half of our cycle.

We can also become super-critical of ourselves, of our surroundings, and of others (this is where the dangerous bit comes in) in this phase. We notice anything and everything that's wrong, which, FYI, can actually be a gift if used correctly. If you own a business, for example, this is an amazing time to look at web copy/reports/figures with a critical eye and to look over contracts too. If there's any editing to do on a book, or presentation, you'll be in the most perfect position to go over everything with a fine-tooth comb, because detail becomes your THING during premenstruation.

Also, if you've been trying to tidy or clean at other times in your cycle? Don't bother. THIS is the time for decluttering. Ask yourself: *do I need these things? Do they serve me? Do they bring me pleasure?* There will be a real desire to release during this part of your cycle, so use it for good!

It can feel hard to connect with ANY benefits in this phase as there's an expectation that we need to maintain the same amount of energy and vitality throughout the whole cycle; it's labeled "weak" or "indulgent" to slow down and be more intuitive, which is most definitely our true nature at this time. When we fight this, perhaps by trying to put in the same amount of hours at work as we did during the pre-ovulation and ovulation phases, or perhaps using coffee and stimulants to keep us going, we're fighting our natural rhythms. It's little wonder our cycles become disruptive, and pain shows up to physically try and slow us down.

If we *do* slow down and connect with the Wise, Wild Woman, our true nature, we discover that we can see and know what's wrong intuitively.

Sensual

Sound the sirens, because if you're in a relationship, this is the part of your cycle where all your STUFF will come up.

Remember back in the pre-ovulation section, I mentioned that if you're putting up with people, situations, and scenarios, and/or pushing certain things to one side, that they'll come back in the second half of your cycle? That time is now.

Yep, the closer you get to your bleed, the more intense these issues will feel. You may experience anxiety, stress, and anger, and if you don't recognize what this is, you may think you're going crazy. Understand, honor, and know that anywhere that you don't feel honored, supported, loved, appreciated, or cared for will *all* show up in this phase.

After much, much experience of allowing my Wise, Wild Woman to fully rant at my husband and not getting the results I'd been hoping for, I'd suggest maybe *not* word-vomiting all your thoughts and feelings that come through in this phase, and instead, write them down. Allow the anger to work through the words before attempting to share any of your thoughts. In fact, I'd recommend having any type of big conversation,G whether it's at work or at home, in the first half of the next cycle when you've had a chance to let it be seen in this phase and bled on during menstruation. This is for two reasons:

1. Our ability to articulate with clarity and from logic is not so present in the second half of our cycle, and this can feel so frustrating and lead to bigger arguments.

2. We might want to really sort it out, yet we may also really want to be alone, so potentially could say something we might later regret—we become *quite* the contradiction at premenstruation.

I used to literally think I was mad before I fully understood my cyclic nature. I thought I was mentally ill. In fact, a previous partner suggested I was and marched me to the doctor. I was a walking, talking contradiction—I wanted to gouge his eyes out with a spoon *and* be loved and hugged by him all at the same time. So, to avoid potential explosions with those you love and live with, give them some insider information: they can boost your serotonin with a BIG hug (but not for too long—ha!), by running you a bath, and by making sure you have an indefinite supply of chocolate.

Along with your hormones, sexual appetite can drop during the premenstruation phase too. Orgasms can take a whole lot longer to reach in this phase than in any other, AND you need more lubrication—ouch-y—AND your nipples and clitoris need more stimulation. All this can leave you questioning whether it's *actually* worth it. The good news is that if you take it slow, it's *definitely* worth it. The endorphins released during arousal can boost estrogen, which can improve mood and relieve PMS symptoms. Hurrah!

Sensational

S l o o o o o o w it down.

Insanity workouts or cardio routines are no good in this phase because, honestly? You'll be far more likely to reach for a chocolate bar and need a nap straight after. Consider some nourishing restorative Yin Yoga instead, sign up to an online IN-YOUR-BODY-MENT class, or maybe Tai Chi—anything that means your movement is more nourishing and nurturing and less hard-core.

Previously, in the last two phases of your cycle, you've been nourishing and tending to the outside world with your ideas, work, passion, and output, and now it's time to nourish your inner world—this phase is an invitation to become much more interested in what needs nourishment and attention internally. What's going on with *you*? What do *you* need? When I mention this to clients, their instant reaction is to declare self-inquiry selfish. It isn't selfish, it's a necessity.

Getting to know yourself—your needs, your wants, and your desires—is what your Wise, Wild Woman craves most, yet so many of us deny her. By doing that, there is a good chance she will demand your attention by kicking you in the ovaries and causing some kind of menstrual-related pain, discomfort, or dis-ease.

This is why I'm SO passionate about encouraging you to chart your cycle. When we know ourselves better, we can show up for ourselves better, and usually, particularly in this part of your cycle, that involves doing less and giving your body, mind, and spirit what it needs the most.

Now, this phase can bring with it carb, salt, and sugar cravings. If you socialized lots during ovulation, this is a great time to get back to basics with your body—don't fret, chocolate is obviously still allowed. In fact, it's actively encouraged. I LOVE chocolate. Unfortunately, my chocolate of choice used to be the kind that came in a purple wrapper and was made up in its entirety of fat and sugar and very little actual chocolate at all. Which is why it's no surprise that actual chocolate, cacao, has now become one of my biggest teachers and medicine, because, really, I can't talk about blood and periods and premenstruation and NOT talk about chocolate, can I? In this part of my cycle specifically, I use cacao as a way to call myself home and to remind myself of who I was before I forgot.

Ixcacao—which literally translates as SHE Chocolate—is the goddess of the brown, gooey yummy-ness. She's a rich, sexy, and abundant earth goddess, and she's ALL powerful. She's feminine and fierce, and from my experience, working with her will give you what you need—not necessarily what you think you want.

So, every time I enter my premenstrual phase, I take her medicine. It's not processed. It's raw ceremonial cacao, and it's bitter. I melt it down, and how I'm feeling will dictate the ingredients I choose to mix with it—sometimes chili if I need to evoke some fire, and sometimes coconut milk and honey if I'm in need of comfort. I then enter into a divinely delicious relationship with SHE, the cacao deva. I allow her to take me right down deep into my womb and unlock my feminine intuition, my knowing, my inner priestess.

I sometimes host SHE chocolate experiences that are four to five hours of bone-deep nourishment. These experiences invite women to enter into a self-devotional practice, to meet their Wise, Wild Woman, the one who craves to go deep within, the one who demands personal time, away from family and responsibilities, to come home to herself. It's powerful and potent SHE medicine, and I LOVE to share it.

However you choose to commune with your Wise, Wild Woman—sacred chocolate, drumbeats, movement, deep yogic sleep—just know that you do NEED to connect with her. Each cycle, you NEED to invite her in with loving arms. You NEED to let her take you by the hand and lead you through the messy, raw, and vulnerable parts of who you are and shine a light on them so you can know them, and get down and dirty and mud wrestle with them.

For me, one of my shadows during the premenstrual phase has been binge eating. I thought I was just out-and-out greedy for a long time, but it wasn't until I started charting my cycle that I realized my appetite was significantly larger at premenstruation. I was then able to see that, yes, I was greedy, but only for one week out of four—and the reason I binged? Because I wasn't looking after myself during the premenstruation phase. I'd ignore the signs my body was giving me to slow down, and instead work hard, stress out, create a metric shit ton of cortisol, and then spend the next six to seven days craving chips. And Dairy Milk. And pizza. And... you get the picture, right? I KNOW this isn't just my story, which is why I've asked **Ani Richardson**, nutritionist, and the author of the book *Love or Diet*, to share her insight and knowledge on premenstrual nutrition.

PREMENSTRUAL NUTRITION

Do you find that your urge to eat becomes stronger during the week before your bleed (days 21 through to 27 of your cycle if you are on a 28-day cycle)? Do you find that you crave specific foods? For me personally, I find myself with an increased appetite at this time and a real wanting for carbohydrate-rich foods–this is a common finding and widely backed up by medical data.

If you're not yet friendly with your cycle or you have emotional eating or other eating issues, then this time can be an uncomfortable or frightening time. When you're eating, you may feel "out of control."

Take a big, deep breath. As you come to chart and know your cycle and appetites, you can prepare and feel comfortable at this time, safe in the knowledge that your changing appetites are "natural."

There's a lot of scientific research that can help to reassure us that the hunger we experience prior to our bleed is a natural part of our cyclical being-ness. There's no need for us to be self-critical or judgmental about our eating patterns. Instead, we can be wise and savvy.

Around day 20/21 of the menstrual cycle, there's a peak in our progesterone levels and a secondary peak in our estrogen levels. This time in our cycle is called the mid-luteal phase. Or we could just call it the premenstrual phase. Scientific studies have found that these changes in our ovarian hormones can predict changes in appetite, consumption, and emotional eating in all women, those with and without eating issues/disordered eating patterns. The studies suggest that in all women, emotional eating scores are highest during this premenstrual, mid-luteal phase. This is, as we know, a time when most of us feel heightened emotions anyway.

IMPORTANT NOTE: *This is not the time to begin restricting eating or attempting to "diet"–there is evidence that women who try to restrain their eating with dieting may be more impacted by the urge to eat during the premenstrual time.*

It's important that we get to know and embrace our cyclical appetites, that we understand that they're natural and nothing to be afraid of. By getting comfortable and familiar with our hunger, we can be prepared with food, self-love, and nurturance at this time.

It's important that we don't feel guilty about this naturally increased premenstrual appetite. Some studies suggest that total food/calorie intake is significantly higher in the premenstrual phase compared to the rest of our cycle. Over the course of a cycle, we naturally balance out our eating (we're pretty awesome like that). Attempting to restrain ourselves harshly at this time could lead to binging. This is a time to be gentle with yourself and listen to your deepest wisdom.

Listen to your hunger–is it emotional or physical? How do you feel? What do you need? Are you tired and in need of a rest? Are you happy and wanting to share your joy? Are you angry and do you need to write out the anger, or dance to shake it off?

Are you sad and in need of a good cry? If we're not dealing with our premenstrual emotions, then we're more likely to have cravings or over-eat as a way to stuff down our problems and stuff down the emotions that we're not allowing ourselves to FEEL.

If you do feel physically hungry, then knowing about food and blood sugar levels can be very helpful. It can be helpful to know what foods to choose to prevent further cravings, binges, or increased anxiety. Studies have found that stabilizing blood sugar levels by eating low glycemic index (low GI) foods, which don't spike your blood sugar or disrupt insulin, can actually reduce the severity of PMS symptoms.

Food wisdom

I am an advocate of REAL food. This means eating nutrient-rich, unprocessed foods. Eat a VARIETY of colorful vegetables and fruits (especially berries which are low in sugar). Don't stick to what you know, look at all the different vegetables available, and don't be afraid to try new things. The internet is full of recipes, so if you don't know what to do with a certain item, then Google it!

Include sources of protein and fats in all your meals, as this helps to satisfy hunger and keep blood sugar levels stable. Think unprocessed meats and fish (especially oily fish which are rich sources of long-chain omega 3 fats), eggs, nuts and seeds, avocado, beans, and pulses. Use olive oil, and don't be afraid to try coconut oil or cacao butter in your cooking too.

When you choose and eat food, do so from a place of extreme self-love. Listen to your inner wisdom, and choose food from the heart: nourishing, nurturing foods. Love yourself as you find peace with food. Ask yourself what you would choose to eat if you really loved *yourself at that moment–and if the authentic answer is chocolate, then eat the chocolate!*

I'm totally aware that even talking about food can bring up all sorts of "stuff" for so many of us. In fact, while throwing myself into this lady landscape exploration, there was a vital part of my cycle awareness that I kept ignoring.

While my endometriosis pain had significantly lessened through charting and actively experiencing each phase of the cycle (as I'm sharing here), I still struggled with my weight. I spoke with Alisa Vitti, a hormonal health expert, and she described the amazing benefits I'd experience if I understood the foods to eat for menstrual dis-ease. Now, while I knew that to fully understand endometriosis and why it was happening I'd have to actually *go there*, if I'm honest, the idea of finding out felt a little bit too science-y and was probably going to push a few too many buttons.

The awesome news is that Alisa made this information accessible, and it's this teaching that has helped me gain a complete understanding of how eating in sync with my cycle can and will– if you're willing to do the work–change your cycle for the better AND ease menstrual pain. When I discovered that my craving for carbs during my premenstrual phase was an actual scientifically proven thing, and not just me being greedy, I was then able to see how I also felt less hungry in my pre-ovulation phase. I then found that, actually, if I ate in tune with my cycle, I'd never need a diet book again ever either—hurrah! I now eat, for the most part, an anti-inflammatory diet, and my menstrual pain is no longer debilitating. It's allowed me to create a deeply cleansing and healing self-love practice that works alongside the creative, spiritual, and psychological energies of the menstrual cycle I'd been working with previously, and has brought me into a full, and really a rather incredible, relationship with my menstrual cycle, my body, and myself.

You

Premenstruation is the phase when everything that may not have been addressed in your life or that has been left undone *will* now rise to the surface.

Don't be scared of this; instead, look at it as a cosmic fire. Any pain, cramps, anger, or irritability that you are experiencing is a signifier for everything you've not addressed, things that haven't come to light, or that you've tried to push out of sight or ignore, consciously or unconsciously, because you were so busy "doing" life during the pre-ovulation and ovulation phases of your cycle. They are now making themselves known in premenstruation and are demanding to be seen.

If you're brand new to this work, you may have a whole lot of "stuff" that you didn't even know you hadn't dealt with, but as you chart, you'll start to see that every month around your premenstrual phase, the same issues may start to arise just before you bleed. These are flashing arrows pointing you to the work you have to do. It might not feel like it, but it's actually a gift. It allows you to unfold and unfurl another part of who you are every month—perhaps you're always mad at your partner, perhaps you're pissed off with them because you have issues that YOU haven't dealt with: this is where you pull out that journal and heart riff. Don't analyze or criticize yourself, just honor your feelings as they are and write them down. Write until you can't write anymore, and you'll find that, it will potentially release some of the internal pressure that's happening within you. It will create space. Talk to your partner about what you're doing: let them know that you're doing some work around how you flow in each phase of your cycle, tell them it's so you can stop fighting yourself, warn them that this may mean you handle things a little differently, and ask them to be patient with you.

The stuff that comes up may have a lot of power behind it, a lot of intensity, so be kind to yourself as you let it unfold. Let it tell you its story so you can work with it and not against it—this is gold. It really is the treasure of every cycle, but it's also work that can be deeply uncomfy too.

I recommend spending a whole lot less time on social media in this phase, mainly because as your gaze turns inward, your sense of self is heightened, and you're capable of being hyper-critical of yourself and of others. Premenstruation + social media = an instant in-road for your inner critic to pull you apart. The "she's better than me" and the "I wish I was like her" comments will fly thick and fast, leaving your self-worth feeling pretty non-existent. Your critic can, and should, be used for good. My inner critic is called Thelma, and I like to give her jobs to do that involve meticulous detail—if I'm editing books, I'll always wait until my premenstrual phase because not a spelling mistake or grammatical error can get past Thelma. The inner critic is a topic all of its own, but for now, just know that deep down, she's really just trying to protect you.

How to get the most out of premenstruation

Get real: I love this phase because it's an opportunity for spiritual insight, straight-talking, and truth-speaking. If it doesn't feel good in your heart and body, particularly in this phase of your cycle, don't do it.

Get curious: Don't be afraid to use your psychic insight and inner knowing to feel and see what's REALLY going on. Those things that you might have previously pushed under the carpet… why are they showing up? Use this phase to get curious and seek the truth. Your truth.

Be critical: You can use this phase to clean up, sort out, and edit in business, relationships, and in life. The great news is that you get the opportunity to use this phase, every month, to conduct a life audit/edit. If you do, your menstrual cycle becomes an ever-evolving self-care system that is able to release, renew, and reset every month.

Warning!

There are some shadows to premenstruation that you might want to look out for each month.

If you're stressed, anxious, and pushing yourself to "do" instead of "be," it's much easier to lose your shit in this phase. In fact, if you're trying to push through the second half of your cycle, you may feel overwhelmed. If you need to release the feels, go ahead. Because if you push these feelings down or to one side, they can potentially create physical pain in your body.

It's also when your tongue is at its sharpest. Yes, truth talking is a good thing, but whenever you can during your premenstrual phase, take a breath before you speak. This isn't self-censorship, this is discernment. I speak from experience when I say, some of the worst arguments I've had, and some of the biggest trouble I've got into, have been during my premenstrual phase. Now I work really hard to respond, and not to react, to people/ situations during this part of my cycle.

 *Rituals*

PMS—WTF!

In the days just before you are due to bleed, you may start to feel highly sensitive, and like everyone is on a mission to purposely disagree with you. (Which isn't difficult, what with you changing your mind on what feels like a moment-to-moment basis.)

Premenstrual Syndrome (PMS) is a term used to describe any symptoms that occur after ovulation and disappear almost as soon as your bleed arrives. These can include: mood swings, anxiety, acne, weight gain, crying, sugar cravings, bloating, constipation, tiredness, and between 150 to 200 other varying symptoms if the notes from my clients are anything to go by.

The biggest problem that my clients seem to experience each cycle is not the physical manifestations of PMS, but what feels like Jekyll and Hyde-like personality changes. One client, Dee, says that she knows she thinks and feels differently to the point of being irrational, but that she has no control over those changes. Another client, Ruth, views the world completely negatively: "Everything seems black," it all feels like doom and gloom, and she finds herself crying for no reason. Except there's a reason for Ruth's negativity, and there's a reason why Dee experiences what feels like a personality change.

I believe PMS is an internal playback of the past days, weeks, and even years, of the things that need to be released, dealt with, and healed within you. It can be as simple as you need to slow down and rest—so your body will give you a migraine so that you HAVE to stop and rest. It may be a massive loud-speaker announcement screaming at you to make some big-ass changes in your life. Only you know the answer to that, and you'll only ever find out if you're willing to do the freakin' work.

The pain, the discomfort, the bitchiness, the anger, the frustration is all an opportunity to become aware of what's not cool in your life and to edit and release accordingly each cycle.

As you begin to listen to and learn the stories, signs, and signals of your body and you begin to pay attention to the messages it's trying to share with you, PMS will become a direct-messaging service of innate body wisdom and no longer be something that you fear.

SHE STORY

Connecting with the feminine, by Meggie Hiley

I have recently realized that throughout my life, I've been in conflict between my "feminine" and "masculine" sides. My "feminine" side is more intuitive, more focused on connection, subject to cyclical energies, and spiritual; whereas my "masculine" side is more rational, analytical, focused on outward presentation, and sees progress as a constant, linear forward movement. (I put "feminine" and "masculine" in inverted commas to signal my awareness that they are, to a certain extent, cultural stereotypes that apply equally to both men and women.)

All my life I've done "masculine" things to earn money and recognition, and "feminine" things as hobbies, spare-time activities, and creative projects carried out for their own sake rather than for financial gain or social approval. This means, however, that they often fall by the wayside when things get stressful, as they inevitably do. When this happens, I feel that my "feminine," creative, spiritual side is being stifled and needs to break free—this is sometimes a very visceral sensation of a kind of twisting heat in my womb, belly, and heart (or the Svadhisthana, Manipura, and Anahata chakras, for those interested in chakra energy). I have attempted to address this in various ways: regular yoga practice to tune in with my body, meditation, morning pages, and journaling. One thing that has been key for me is learning to "go with the flow" (literally!) of my cycle—allowing myself to feel the ebb and flow of energy, being aware of which times of the month I feel social and outgoing, at which times I'd rather just be left the hell alone, and at which times I sink into what I can only call a trance-like inward focus. When I allow myself to just give over to this cycle, everything seems to come together—I feel inspired, happy, spiritually connected. But it's not easy to give over. In fact, it's bloody hard!

All of my life, I've been rewarded for pursuing "masculine" goals in a "masculine" way; and even though I'm self-employed and thus theoretically in charge of when and how I work, I find that I'm often my own worst enemy. I repeatedly fall into the trap of taking on work when I should really be tuning in to the fall/winter of my cycle, thinking I need to "set targets," "pursue goals," and "think big" for my business to develop and progress. And so I end up feeling completely overwhelmed by demands that, although they have come from my own brain, feel quite foreign to me.

In short, I fall into the trap of thinking my work needs to be "masculine"—with the same results again and again: exhaustion, overwhelm, and alienation. It's not pretty. This has even had some pretty dramatic physical manifestations: one month recently, when I found myself completely swamped in work and work-related demands (even though I really just wanted to sit in peace and quiet and feel my body getting ready to bleed), one of my breasts became really swollen, lumpy, and inflamed, which was actually really scary (my doctor sent me to the specialist clinic). It clearly was a sign to stop ignoring my body, and my physical and spiritual needs! The ironic thing about this is, of course, that if I allow myself to tune in with my cycle, I work better and in a more focused manner during my spring/summer phases, and gain deeper insights and new ideas during my fall/winter stages, and then emerge into spring feeling rejuvenated and with more energy to get things done. It's a real virtuous cycle, and my body knows it—it's just a matter of persuading my mind that actually, this is an acceptable (or in fact superior) way of doing things. So, if I attune to my body and my cycle, I feel more connected, more energetic, and produce better work both in professional and creative ventures; if I ignore my cycle, I get overwhelmed, tired, frustrated, and—in the worst-case—physically ill. So now, every month, I vow to attune to my cycle and go with my flow. Of course, it doesn't always work. But the wonderful thing about bleeding is that each month we get another chance, and so I know that there will be many months and cycles to come, and I can keep working on this for the rest of my menstruating life (and beyond).

PMS Smoothie

I HAVE to share this yummy combination of good-for-you food, originally shared with me by the gorgeous Leonie Dawson, because it's my go-to morning breakfast during my premenstruation phase. It tastes amazing and, most importantly, it's easy to make AND nourishing.

Ingredients

3oz vanilla Greek yogurt

½ frozen banana, peeled

½ avocado, peeled and stoned

½ cup spinach

1tbsp raw cacao

½ cup chocolate almond or coconut milk

1/8 cup raw cashews

A few ice cubes

½ cup coconut water

3 dark chocolate chips (garnish)

Add the ingredients to a blender (save some cashews for topping) and mix until smooth. Pour into a glass. Chop the remaining cashews and chocolate chips, sprinkle on top, and enjoy. See? I told you it was easy!

In-your-body-ment

What I love to do most during this premenstrual phase (besides munching on chocolate) is to experience some really good, nourishing body squeezes, which is why a seated twist offers many gifts: a stretch in the outer hip and thighs, opening in the shoulders and chest, and a healthy lengthening of your spine.

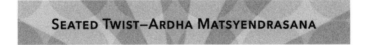

SEATED TWIST—ARDHA MATSYENDRASANA

The twists cleanse and flush the lymphatic system, bringing rejuvenation into the pelvic bowl ready for menstruation.

1. *Start by sitting with legs extended in front of you. Cross your left leg over your right and place your left foot flat next to your right knee.*

2. *Bring your right heel as near to your left hip as you can. Don't push it, keep it comfy. I have a tilted pelvis, so this is a little harder for me. If it's hard for you, simply adapt it until it feels good—that's the golden rule with SHE Flow yoga.*

3. *Put your left hand on the floor by your left hip. Take in a big deep breath, wrap your right arm around your left knee and twist your torso to the left.*

4. *Hold for 10 to 15 deep breaths, or until it starts to feel uncomfy—whichever comes first.*

5. *Unwind slowly, before returning to the start. Switch sides and then repeat.*

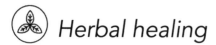 *Herbal healing*

Now is the time to balance any wayward hormones that might feel too extreme. It's also the time to work deeply with your inner mystic wild woman! Keeping blood sugar levels stable is essential, so reaching for nutrient-dense and slow-release foods will support energy levels and emotions. These herbs can help with hormonal balance and have adaptogen qualities (increase resilience to stress and change).

Shatavari (Asparagus racemosus)—see Chapter 4

Vitex (Vitex agnus-castus)—see Chapter 5

Raw cacao (Theobroma cacao)

Uplifts the soul, improves circulation, increases concentration, and so much more. It contains a treasure trove of nutrients; the most important for women's wellbeing are magnesium, iron, anandamide (the bliss chemical), PEA (the love chemical), and the feel-good neurotransmitters serotonin and tryptophan—all affirming those premenstrual cravings... our bodies are so wise!

Ashwagandha (Withania somnifera)

A wonderfully restorative tonic. One of the best rejuvenating herbs for the mind, upon which it has a nourishing, calming, and clarifying effect. It promotes deep and restful sleep, regenerates the hormonal system, and is a tonic for reproductive tissue. It's good for weak pregnant women as it's said to stabilize the embryo. Like all good tonics, it's known to be an aphrodisiac, recommended to increase sexual energy and enhance fertility.

 *Flower power*

Yellow rose

Keywords—Helps with self-doubt, fear of getting "it" wrong, feeling depressed, and self-criticism

The essence of yellow rose helps to lift our spirits when we might be feeling in a funk, sad, irritated, or feeling like what we have to say isn't good enough. It guides us toward letting go of being right and controlling anything we're trying to do. It helps us to put our cynicism aside and brings in the sense of calm, solace, and gratitude. The yellow rose essence is gentle, warm, friendly, uplifting—and if you're feeling none of those, she will reconnect you.

CHAPTER 7

Menstruation—Let It Flow

Female archetype: Crone

Moon phase: Dark/New Moon

Season: Winter

Element: Earth

Tarot card: The Moon

Crystals: Amethyst—aids spiritual connection

Ruby/garnet—cleanses negative emotions in blood

Essential oils: Clary sage—soothes muscle cramps, restores emotional wellbeing

Nutmeg—warms, stimulates and elevates awareness

Vanilla—provides deep calm in the darkness

Mantra: "Let it flow."

Song: *Time to Flow*—D–Nice feat. Treach

Perfume: Ginsberg is God from Bella Freld—as comfortable as old books and cashmere

The science bit

If you're not pregnant, you'll be bleeding.

The day you start to bleed is day 1 of your cycle, and this is the day to start charting if you haven't already. Depending on your cycle, your bleed can last anything from between three to eight or nine days, and varies from person to person. The first few days of your cycle may leave you feeling achy and tired, but from day 3 onward, your estrogen levels will start to rise again, and you'll experience a boost in your energy, mood, optimism, *and* your brain skills. When estrogen rises, it also boosts your levels of testosterone, and when *this* increases, it amps up your confidence, self-esteem, and courage. But don't get too excited on the first days of your bleed as these happy hormones, while rising, are not high yet. In fact, in the first three or four days of your bleed, they are staying low so that you rest and immerse in self-care and preservation.

Let it flow

In pre-ovulation, we met the Maiden, in ovulation the Mumma/ Creatrix, in premenstruation the Wise, Wild Woman, and in menstruation, we meet the Crone.

The Crone understands that without death, release, and letting go, there can be no fresh starts, no new beginnings—and that's what our menstrual bleed is, a chance for us to die each month so that we can be reborn. Yep, our menstrual cycle is a built-in self-renewal system. In yoga, we talk a lot about the death of the ego, and it's as if during our pre-ovulation and ovulation phases, our ego, our sense of identity, builds, then, as we step into premenstruation and menstruation, if we're working with our cycle, we let that go. We die to what we thought we knew, to what no longer serves us, so that we get the chance to start over, taking with us the lessons and the teachings of the previous cycles. EVERY MONTH.

The Crone is old, and she is wise. She represents the woman who is now in menopause. She no longer bleeds; she doesn't need to. She knows that our bleeding years are when we unfold and unfurl, when we learn and collect information about ourselves, others, and the world around us, so that when we do reach menopause, we can truly own who we are and hold ourselves sovereign. We no longer have to experience each phase of the menstrual cycle, because when we become the Crone, we have all the phases within us, and we have an all-seeing perspective.

The Crone is immune to the bullshit, to the many issues that consume young women—the constraints of partnership, beauty, social conditioning, the need to fit in, whether to have children or not. The Crone *should* now be able to live a life free of these restraints; except in modern society, so many women enter menopause without any preparation or knowledge of the power they hold there and feel "washed-up" and like they're surplus to requirement.

The reason I share my work is to help women recognize the power of each stage of womanhood as they meet her in each phase of their menstrual cycle. This way, when you no longer bleed, when you move from the Wild, Wise Woman to Crone, you know the potency of who you are. You know the wisdom you hold. You know your worth. You know you're sovereign. You drop the pretensions, rules, and limitations that current society has you live by and recognize the Crone as compassion AND fierceness, as understanding AND truth, as caring AND also giving zero fucks.

Look for representations of these women in your life. It won't be easy as the media and society are even less fond of the Crone than they are of the Wild, Wise Woman—but they *are* out there, and when you meet them, they are bloody glorious. They see the truth everywhere, and at menstruation, this is your opportunity to step into your Crone archetype, to see the truth in your life, in your relationships, and in your actions.

If you experience pain, discomfort, or a total aversion to this phase of your cycle, then I invite you to explore your relationship with the Crone. Get still with yourself, set a timer for fifteen minutes, and write your freakin' heart out. Ask yourself how you feel about getting old; does it scare you? What does the idea of dying conjure up? Fear, darkness, regret? How we respond to this phase of our cycle can sometimes be an indicator as to how we feel about stepping into the Crone archetype in our life. I actually really look forward to bleeding, which I appreciate may sound like a very strange thing to admit especially when I've previously suffered incredible, debilitating pain during my bleed, but for me, it's become a time that I get to go deep. SO deep.

I'm a Scorpio, so I love to go as deep as the ocean and delve in the dark, and the Crone loves to take us there. She's not afraid of death—she stares it straight in the eyes, and at each bleed, I experience it too. A death to who I previously thought I was, as I strip layer after layer away each cycle.

Every month, like the serpent, we shed. Our cycle gives us this incredible opportunity to shed parts of ourselves that no longer serve us, clearing space for something new to be created.

That new thing won't be the same as what was lost, because each cycle, if you work with it, will give you the opportunity to learn new lessons about yourself, to release what no longer works, and to create a new and amazing life. EVERY FREAKING MONTH this is happening, have I mentioned that?!

This phase is represented by the Dark Moon; it's midwinter, the pagan Sabbat of Yule, the dark before the light, that darkest moment—and we've all experienced those, right? The dark nights of the soul, the kind that brings you to your knees, where you never think you're going to experience light again. Yet we celebrate this cyclic winter solstice, the shortest amount of light each year, because we know, with total certainty that the light IS returning. This is what happens with our cycle in the days before we bleed. It calls us toward the dark, it calls us to be willing to let go of what no longer serves us, and when we bleed, there's both release and relief as we cycle toward the light of ovulation— midsummer—once again.

Your menstruation SHE powers

Each phase has a set of superpowers you can access and hone to make life a whole lot sweeter during this part of your cycle:

- You become super-clear about your purpose, why you're here, and what matters. If I'm asked to make a decision during premenstruation, if it's not urgent, I'll wait until I bleed because, with the release of the blood, everything just makes more sense.

- You're able to connect to SHE/universe/source directly– obviously, this is possible at any time, but during menstruation, it's as if you have a direct line: dreams are more vivid, and if you're open, you basically become a divination rod, for insight, messages, and wisdom.

- You can identify and release the stuff that no longer serves you really easily. Things that got you angry and frustrated during premenstruation no longer hold their power here, so you're less likely to feel attached to anyone or anything.

Yep, there's a five to eight-ish day period of time each cycle, from the first day you bleed through to the day you stop bleeding, when you are witch/sorceress/woman.

This has probably been the biggest revelation I've discovered since charting and working with my cycle: if I totally honor the menstruation phase, I'm given so many signs and messages into how to move forward with a situation, and which direction to take in my life, that I no longer worry about what I *should* be doing, or where I *should* be going. If I have a big-ass decision to make, I simply bleed on it. Meaning, I wait until my bleed time, and then I stop, look, and listen for signs as to whether it's something that I need to do in my life. Sometimes when people ask me as to whether I'd like to take up a new role or be part of a project or collaboration, I'll tell them to let me bleed on it, and I'll get back to them straight after. This is how I do business, and it's how I do life.

I bleed on it.

How to activate your menstruation SHE Powers

You have access to these particular SHE powers during the menstruation phase of your menstrual cycle EVERY. SINGLE. MONTH.

The best way I've found to fully activate them and use them to their full potential is by aligning your SASSY–Spiritual, Authentic, Sensual, Sensational YOU–but, remember, it's up to *you* how you choose to use and interpret the information in each phase. This is just simply a guide to help you on your way, OK?

Spiritual

Many women, most of us, in fact, have been led to believe that their bleedtime is shameful or embarrassing. They whisper about it in hushed tones. I have friends that won't even talk about my work because they're mortified that I talk so openly about blood and periods and menstruation. I have clients that cringe as they tell me about their monthly cycle because they fear I'm going to make some kind of judgment because they bleed, while others ask if what they're sharing with me in sessions is too much information. FYI: you can NEVER share too much information about your cycle and bleed with someone who talks about vaginas, yonis, menstruation, and ovaries EVERY day. I LOVE understanding how different it is for each of us and finding the clues as to what our body is trying to tell us through the issues that we face. For the most part, unless you're really lucky and have a clued-up mumma, this message of shame has been passed down to us by our mummas and grandmummas. So, right now, I'm going to invite you to open up to the idea that menstruation, our bleedtime, is sacred.

I want you to open up to the idea that this time is important. Not just for you, but for every person and life that you touch or come into contact with. One of the ancient mysteries about bleedtime is that it's created as a space where the feminine cleanses not just herself, not just the physical where we release the lining of the uterus that's no longer needed, but a cleansing of the issues of our families and our communities, too.

See? You're literally SO powerful and potent at this time, and it's the spiritual work that you do during your bleed when you're in a place of total consciousness and awareness, that can literally change the world.

Many native and indigenous people knew this. The reason we don't is that this mystery is passed from woman to woman. It's something that has to be opened up to you by another woman—just like we're doing here in these pages together—and you have to choose to receive it, or not, into your own life and life experience.

That's why during your premenstrual phase, all that "stuff" came up. The "stuff" that you could no longer hold on to, that needed to be aired and addressed. It came up so that you could release it. You can cleanse your relationships, your business, and your life—you no longer have to hold onto that "stuff," you can forgive and let go. That's why during our bleedtime, our ancestors used to go to what some people refer to as the red tent. If you haven't read *The Red Tent* by Anita Daimant, you NEED to—I think it should be required reading for every woman. The red tent was a space for women to bleed together and to rest. Since the release of Anita's book, there has been an emergence of red tents across the globe. Some of you may be lucky enough to have one in your area, a space where you can gather with women each month, but there's not always one we can go to when we bleed, so I make my own.

Literally, sometimes, I'll make a red blanket tent fort and curl up underneath it. You don't need an *actual* red tent though: you could find your own red blanket that you snuggle under at bleedtime, buy a red journal and red pen to write down the downloads you'll receive during this conscious time (and believe me, if you're open to it, those downloads will come thick and fast), tie your hair in a red scarf, have a bracelet or ring that you dedicate to your bleed (I have a red garnet ring that I wear when I bleed) to remind you that you're in your red tent, even if you have to go to work or be out in the world.

What do you do in a red tent?

Slow down and allow yourself to rest.

If possible, give yourself the first day of your bleed off. For some, this will sound ridiculously indulgent, and for those of you with children, careers, and families, especially, I understand that an entire day out may just not be possible. At every SHE chocolate experience, I host, once women have drunk the sacred cacao, the afternoon is then spent in a state of relaxation—Yoga Nidra, soundscapes, journaling, nesting, sleeping if needed. It's an invitation to take a breath, yet the one thing I hear repeatedly is: "that felt SO good, but I feel so guilty for taking time for myself."

This is bullshit.

You NEED to take time for yourself in order to be fully present for your family, those you love and work with, and, most importantly, for yourself. When possible, I give myself the first day of my bleed off, and because the Viking works super closely with my cycles too, he honors me, and he'll do all the cooking and give me space. I honor the shit out of my bleed because I know how powerful and potent it is. This first day is when you are most likely to receive visions, downloads, and dreams that have meaning, insight, and revelations that are clearer than at any other time of the month, and I like to be totally ready for those. I *can* schedule my work accordingly, I don't have children, so it *is* easier for me to make time and space, but if you can't take a day, take a morning. If you can't take a morning, take an hour. If you can't take an hour, take 15 minutes. The important thing is that you don't let your bleed simply pass you by each cycle, that you don't try to "keep going" and maintain the masculine norm, and that you DO take time to give your bleed a li'l wink of recognition—because when you honor your bleed, you honor SHE, and when you honor SHE, you honor yourself.

Authentic

During this phase, you're not going to want to focus much on the details and actualities of work, or of anything, really, but you *will* receive ideas, downloads, and a heads up on areas that need attention. If there's something at work, or in your life where you're feeling dissatisfied, that WILL come up, especially in the first few days of your bleed. After that, from day 3 onward, you'll start to think: *What do I want to start to build? What do I want to begin?* Ideas and inspiration will come thick and fast—but don't act on them, not right now. Wait till day 7 or 8, when you're moving into pre-ovulation to begin *that* process. For now, rest, receive, and just write them down.

If you do have to work, make this the time when you set clear boundaries.

Be super-clear about what is and what isn't possible. Despite what those pesky "feminine hygiene product" adverts (don't even get me started on the use of the word "hygiene") will have you believe, you really do *not* need to be riding a rollercoaster, sailing a boat or roller skating along a beach promenade in white trousers when you bleed. It's in this phase you'll want to send me kisses and chocolates (feel free, I'm always very happy to receive) for encouraging you to do that batch cooking during your ovulation phase, because right now, all you'll want to do is to simplify and slow down. I know this may be a bit of a shock to your family, especially if you have a family that's used to you being that do-it-all superwoman who does everything for them on demand. Talk to them about what you're exploring in the pages of this book. Tell them that you're going to slow down while you bleed, explain to them why, and, most importantly, if you feel guilty about taking time out, acknowledge it, but be sure to persevere.

So many of my clients, on hearing me suggest that they slow down and take an afternoon off when they're bleeding, look at me with total terror: "But I've got a family to feed/look after. I'm so busy, how will I find the time to do nothing?" I get it, I really do, but it's this constant need to "do" for the entire cycle that's causing women to suffer pain and menstrual dis-ease. Women find the idea of doing nothing so bloody difficult, but your only job to do when you're bleeding is to recognize that you're doing the most work by doing no work. Your body, your relationships, and your family *will* thank you for taking the time to do nothing, especially if you're someone that experiences pain at bleedtime. If you are, this will *usually* be a result of having pushed yourself too hard throughout the other phases of this particular cycle. (Of course, it's not ALWAYS this simple.)

Sensual

I want to get straight onto the sensual act of arousal, because while your body might feel tender and swollen and while you may be experiencing pain and discomfort, your libido is actually turned on from day 1 of menstruation. If you can, and you absolutely should, get past the idea that it's "messy," you'll discover that orgasms aren't only really easy to reach during menstruation, they also release the hormone oxytocin, which is a natural pain reliever AND creates insta feel-good vibes. I would say, though, that because your bleed is such a sacred time, only people who are going to treat you like a priestess and respect what an honor it is to be able to share this powerful time with you should be given permission to bring you to orgasm—I really mean that. I'm talking self-touch in devotion to yourself, or long-term soul mate/life partner sex; the deep, touch-your-soul kinda sex is the *only* kind I'd be entertaining during menstruation.

To honor my bleedtime really deeply, days 1 and 2 are for me and SHE—I even sleep in a separate bed sometimes, and I write, and I sleep, and I eat. But from day 3, I certainly entertain the idea of being totally honored as a sacred, sensual priestess in ALL the ways. Wink.

Sensational

To really reap the rewards of menstruation, I recommend resting, relaxing, and observing as much as possible. Allow tears to come and flow, allow whatever you feel—laughter, tears, tenderness, sensitivity—to come and flow, because when we bleed, we feel it *all* really powerfully. At times, we call it madness (I know I definitely have) or berate ourselves for being a "silly" woman who can't control her emotions. But we're not "silly," we're cyclic, and allowing yourself to feel and experience your "feels," is a cyclic power. The more you feel, the deeper you connect with yourself and get to know yourself. In order to *really* feel it, I recommend doing no exercise on days 1 and 2 of your bleed. That's none, nada, absolutely nothing.

When I first started talking about this, personal trainers HATED me for suggesting there were days in a women's cycle where she shouldn't work out, but now, I'm really blessed to consult with trainers and sportspeople across the world to create workouts that support, and are in sync with, the cyclic nature of women. Personally, I might do a little light stretching on days 1 and 2, then, from day 3 onward, I'd suggest some yummy Yin Yoga or sensual movement. Nourish your body, don't eat *too* much sugar or fat, drink warming tea—I LOVE licorice tea, but rose tea is also super-loving at menstruation too—and allow a little self-expression through writing, art, and chanting.

I'm not one to tell anyone what to do, but personally, I don't use tampons. Why? Because I believe that what we need to let go of—physically, emotionally, spiritually—flows out of us through our blood. Many of us use tampons to pretend we're not bleeding, to "stuff up" or to "absorb" the situation, and this can create pain and discomfort. I prefer handmade cloth pads—I know, I know, I used to really laugh at the idea and wrote it off as something my hippy mates did, but they convinced me to give them a go. They come in really gorgeous colors and materials, and it feels like I'm wearing fluffy knickers. After use, I soak my pads in a bowl of water and give that water back to Mumma Earth before popping them into the wash. I *totally* appreciate how out-there this all sounds, but if you *do* use tampons, and suffer from bad cramps, try a month or two without them, and I can pretty much guarantee the cramps will lessen because the blood is able to flow out and through you.

> ### NOTE
> ....
> *If you're feeling like it's a hard "no" on handwashing your blood out of cloth pads and offering it back to Mother Nature, the amazing, wonderful, and wise Dr. Christiane Northrup wrote in the foreword of my book,* Love Your Lady Landscape: *"Blood is the richest source of stem cells known to humanity. Try using some on your plants and herbs." Just saying.*

Here are some of my favorite alternatives to the disposable, bleached, and mostly plastic, menstrual products currently on offer in our supermarkets.

Cloth pads

These pads are made of cloth and button around your knickers to absorb your flow, just like a disposable pad. You wear, remove, soak in cold water, wash, dry, and use again. They are made of natural fibers, they allow your skin to breathe, and can eliminate rashes and yeast infections.

Cloth pads are a brilliant alternative for those who find tampons uncomfy or who have a tilted uterus (if you do, you can find it tricky to use internal products–I'm most definitely someone who does). I'm not going to lie—I put off trying these for the longest time, but after using bleached white disposable pads for most of my menstruating years, I now have a set of leopard print and velour pads, and my vulva lips are insanely grateful. They no longer itch and now think they're the lady-part equivalent of Cleopatra. Honor Your Flow, Lunapads, and Gladrag pads are some of my personal favorite suppliers, but if you're creative, you can also make your own, and there are lots of patterns on the internet (see also REDsources at the back of the book).

Menstrual cups

Made of silicone or natural gum rubber (latex) that you insert into your vagina, a menstrual cup that collects, rather than absorbs, your menstrual blood. You can insert it in the morning and potentially leave it in for up to 12 hours, depending on your flow. They come with instructions on how to insert them, but basically, the soft cup is folded and inserted into your vagina, and as it unravels, it creates a seal. The seal stops the blood from leaking, and the cup collects the blood. To remove it, you pinch the base of the cup to release the seal and pull gently. You can either pour the blood down the toilet or use it on your plants.

Menstrual cups retain the natural moisture of the vagina and keep your vagina free of icky chemicals. Unlike tampons, they also conform to the natural shape of the vagina, for optimal comfort.

Sea sponges

These are actual sponges from the sea that absorb your menstrual blood. You remove it, rinse it, and reinsert it. A small piece of string can be sewed onto the sponge to help with removal. They are soaked at the end of each cycle for sanitation. They're inexpensive, and they can last for quite a few cycles.

Period underpants

These are underpants/knickers that have a super-absorbent, invisible built-in section that can hold up to four tampons-worth of blood. They draw away wetness, they don't smell, and they're meant to be totally leakproof. I have friends and clients who say that these are a total gamechanger for them.

Just to note, I'm aware that for many, even the menstrual products that you buy in the chemist or from the supermarket are far too expensive, and thank goodness that many places around the globe are now starting to FINALLY lift their taxes on them. I'm sharing the reusable menstrual prods as an alternative if they fit your budget because, yes, they can initially be a little more expensive than buying the throw-away versions, but long-term I believe they *are* worth it. And if all the health benefits aren't enticing enough, you'll save money each month by using reusables, *and* you'll be supporting the planet too. Hurrah.

Interestingly, I found after speaking to many clients that the reason lots of women want to use tampons is so that they don't actually have to touch their blood. I get it, we've been taught that it's dirty and that we should be ashamed and/or embarrassed by it. Don't be afraid to touch or handle your blood. It's part of you. It isn't dirty or something to be ashamed of. It's the very stuff of being a woman. It's the bed of nourishment and goodness that all life starts from. So, I invite you to touch it, to feel its consistency, and, depending on how daring you feel, allowing yourself to bleed freely at night. Take a towel, fold it up, and tuck it between your legs. You might be shouting at me now, saying, "But you don't know how much I bleed!". I get it, I'm a REALLY heavy bleeder—but what happens is that the blood flows into the creases of your vagina wall, and everything that's not caught by your vagina, will spill out onto your towel.

You'll be amazed at how different you feel when you are conscious of your blood and not trying to avoid it or make it clean. Just by doing these practices, my cramps have gone away—cramps are stagnant energy that gets stuck in the pelvic area, and when you use tampons or internal products, that energy that wants to flow out of you with your blood, is being contained, which causes the cramps. I used to bleed heavily for days on end, more than I didn't bleed some months and I was super-heavy, and I had a LOT of pain. But since allowing my blood, and all the emotions that come with it, to free flow as often as possible, the pain is now much less, and I bleed less too. If you experience heavy bleeding, asking you to trust me on this is a big deal, but why not try it? At least for one or two cycles. Learning to be OK with my blood and the fact that I bleed is one of the biggest teachers I've experienced in my personal exploration.

NOTE
....

I'm aware we all experience a different relationship with
menstruation, especially if you have what's called a
"gyno-urinary disorder"–endometriosis, fibroids, PCOS,
basically anything gynecological and pelvic floor related.
I experience both endometriosis and PCOS, and it's
because of them I got curious about my lady landscape.
Have I healed them completely through the practices
I've shared here? No. Are they significantly easier to
manage and to be with since I've acknowledged, honored,
and revered my cyclic nature and my menstrual bleed?
Abso-bloody-lutely.

You

Make the next three months you chart all about you. Every time
you bleed and shed, let go of what no longer serves you. You
can do this through ritual, through journaling, by writing it down
and burning it in a fire at a Dark Moon, or simply by giving it
back to the universe in prayer–whatever works for you. After a
few months, you may start to get stuff that you simply know is
not yours–that's when you know you're doing the work of SHE.
You'll start feeling the echoes and the pain of the women in your
family, or women in your lineage.

That's the power of your bleed. Many women in many cultures
were sent away during their bleedtime, but not because of
shame or because it was dirty–it was because a bleeding woman
had specific work to do.

Honoring and communing with the potency of this practice every cycle is the REAL power and is what we REALLY mean when we talk about being empowered and recognizing the power in your own body. If you're bleeding, you can practice this.

For those going through, or who have been through, the menopause, don't feel like you've missed out on this practice. Use the Moon cycle and set the intention of honoring this work at the Dark Moon. If you have naturally gone through the menopause, then ALL this power is inside you, it's contained within you—you're incredible, lady! The work that the women who are bleeding do, you can access at any point. That's the true power of a menopausal woman and something which isn't often talked about. We consider going through the menopause as meaning we're no longer useful, that we're past our sell-by date, but we couldn't be further from the truth. All the powers that we've talked about, the powers that are unlocking and unfolding in us in each phase, every time we bleed? You contain ALL of them.

So never, EVER, underestimate a menopausal woman. EVER.

How to get the most out of menstruation

All of these points will go TOTALLY against what you've been taught and told to be as a woman, but I want you to at least consider them as you experience your menstruation phase.

Say "no" more often: Is that word even in your vocabulary? Your natural tendency, especially if you're a mumma, may be to give yourself, your time, and your resources freely. Practice saying "No" in your menstruation phase. Don't rush to answer emails and texts. Avoid to-do lists (at least for day 1), and be OK with turning down social engagements if you'd prefer to take a bath or curl up with a book. Setting boundaries is a major stepping-into-your-power practice, so honor that in menstruation. Honor yourself.

Be nice to yourself: Buy yourself a bunch of flowers—check out Sal's recommendation for menstruation's power flower at the end of the chapter. We've a tendency to wear our worst underwear during menstruation because we "don't want to ruin them," but I bought some really pretty underwear that I wear ESPECIALLY for my period.

Practice being gloriously selfish: I dare you. Start with declaring an hour of alone time on the first day of your bleed, then over the next three months, start to declare as many hours as you can on day 1. Make it sacred, make it restful, make it devotional. Or simply sleep. Your call.

Warning!

There are some shadows to menstruation that you might want to look out for each month.

If you don't slow down, or at least lessen your workload, there's a chance you'll experience pain, discomfort, and dis-ease during your next cycles. And if you're a woman who doesn't suffer pain or discomfort during her bleed, yet still does life at 100 miles an hour, please know that while this may not be impacting you now, it may well impact how you experience going into the menopause. Experiencing your cycles fully each month; the ebb and the flow, the masculine and the feminine, the light and the dark allows us to be prepared to enter our Wise, Wild Woman phase of perimenopause in reverence for what has gone before, as a priestess of the ALL-knowing, not kicking and screaming and wishing desperately for our 20s to return.

If you dive deep emotionally during your menstruation, you may get lost and feel a sense of overwhelm at what you may be feeling called to do. But know this, you're made of magic. You're here, reading this book, because you're ready for it. I'm so glad that you are. The world needs "called" women.

> …by honoring the demands of our bleeding, our blood gives us something in return. The crazed bitch from irritation hell recedes. In her place arises a side of ourselves with whom we may not at first be comfortable. She is a vulnerable, highly perceptive genius who can ponder a given issue and take her world by storm. When we're quiet and bleeding, we stumble upon solutions to dilemmas that've been bugging us all month. Inspiration hits and moments of epiphany rumba 'cross de tundra of our senses. In this mode of existence, one does not feel antipathy toward a bodily ritual that so profoundly reinforces our cuntpower.

–Inga Muscio,
Cunt: A Declaration of Independence

 # *Rituals*

Create a sacred space

Whether you call it a red tent, a Moon lodge, or simply "your space," making it sacred at menstruation is an essential part of loving your lady landscape.

I have a menstrual medicine box that I set up as an altar each time I bleed, and it's filled with things that make me feel good at menstruation: my journal and a red inky fountain pen; my *SASSY SHE Oracle Cards*; rose absolute, sandalwood, and frankincense essential oils; crystals (my favorite is a carnelian that looks like a vulva–I bought it in Glastonbury, and it fits perfectly in my palm, so I hold it a lot when I'm bleeding); my goddess statue that's been ritualized with my menstrual blood; runes; and a MAC red nail varnish.

How to create your own sacred space at menstruation:

Create your altar–Find personal items that are meaningful to you. I have my menstrual medicine shoebox, and I also have a little red organza bag that I carry in my handbag to collect things I find on my travels, maybe a shell. My latest addition is a badge with a pair of ovaries on it and the slogan: "Grow a pair"–I love that!

Surround yourself with inspiration–Fill your sacred space with anything that's uplifting and makes you feel good–quotes, good-smelling candles, crystals, symbols that feel supportive.

I have a *Mary is My Homegirl* mood board that sits on my altar, which has pictures of all my favorite Mary's: Lady of Guadalupe, the Black Madonna, Mumma Mary, Mary Magdalene. I also have pictures of my mumma and my nanna to honor my matriarchal line, and I also have a picture of Taylor Swift because I dig her. Whoever and whatever lifts you up, that's the only rule.

Create a ritual–Once you've created your sacred space, it's totally up to you what you do with it. Sometimes I curl up under a red blanket and listen to a Yoga Nidra, sometimes I'll move my body in a slow and sensual way, other times I'll simply light a candle and give a wink of acknowledgment. Whatever you do, make it meaningful to you.

⊛ *In-your-body-ment*

Now, I know I said to do little or no exercise during the first days of your bleed, but this pose is so good that you WILL thank me for sharing it. If you experience any inflammation or constipation at menstruation, this will open up the pelvic bowl and create space. It's also a great stretch if you experience back and thigh pain. Wait until at least day 2 or 3 of menstruation to try it out, and you might want to get a few blankets and fold them up so you have something soft to recline your back onto, rather than just your mat. If you have a bolster cushion, these work great placed along the length of your back.

RECLINING BOUND ANGLE POSE–BADDHA KONASANA

The natural tendency in this pose is to push your knees toward the floor in the belief that this will increase the stretch of the inner thighs and groin. But especially if your groin is tight, pushing down your knees will have just the opposite of the intended effect: the groin will harden, as will your belly and lower back. Instead, imagine that your knees are floating up toward the ceiling and continue settling your groin deep into your pelvis. As your groin drops toward the floor, so will your knees. If this feels uncomfy in any way, simply place lots of cushions or folded up blankets under both your knees and allow them to be held. The aim is to find delicious comfort and ease. Stay here anywhere from 30 seconds to 30 minutes.

1. *Start by sitting with your legs out in front of you.*

2. *Take a deep breath and as you exhale, lower your back toward the floor, first leaning on your hands. Once you're leaning back on your forearms, use your hands to spread the back of your pelvis and release your lower back and backside through your tailbone.*

3. *Bring your torso all the way to the floor or to the cushion, supporting your head and neck on a blanket roll or bolster if needed.*

4. *Grip your topmost thighs with your hands and rotate your inner thighs slowly outward, pressing your outer thighs gently away from the sides of your body.*

5. *Next, bend your knees and let them drop to the ground, with the soles of your feet touching. Shift as much as you need to in order to find a comfy position. Lay your arms on the floor and relax. Ahhhhh.*

6. *To come out of this position (which you'll never want to do), use your hands to press your thighs gently together, then roll over onto one side and push yourself away from the floor and into a seated position.*

 ## Herbal healing

Support your inner winter with herbs that love the womb and soothe feelings of heaviness, dragging, pain, or intense cramping. Love yourself with herbs, womb wraps, and plenty of rest.

Raspberry leaf (Rubus idaeus)

Famous as a pregnancy herb, it supports healthy labor and delivery, relaxing and toning actions with an affinity for the womb. This herb strengthens and tones pelvic and uterine muscles while soothing and relaxing at the same time—magic, right?! Excellent for intense cramps and spasmodic pain, it can be highly astringent during heavy bleeding. It's rich in the minerals calcium and magnesium. Tastes great as a leaf tea with a sprig of lemon balm and honey.

Lady's mantle (Alchemilla mollis)

A powerful female herb for any time during a women's reproductive life. It helps relieve mild aches and pains during menstruation, with a tea or tincture able to stop spotting between periods and lessening excessive menstrual bleeding, helpful for menopausal imbalances.

Cramp bark (Viburnum opulus)

A uterine sedative and tonic, it relieves cramps, spasmodic pain, and uterine irritability, it regulates contractions in labor, and soothes afterpains. It's a specific remedy for pains in the thighs and back and a bearing-down expulsive pain in the uterus.

 *Flower power*

White calla lily

Keywords—Transition, death (not literally), rebirth, letting go, stillness, change, intuition, reflection

The essence of calla lilies gives us the chance to shed the old, to let go, accept, and surrender, and to become something NEW. They show us ways to shed our skin and to recreate who we are. In white, these flowers help us to clear what no longer serves or is needed, and offers support in rebirthing another self, a new version. It makes tapping into intuition much easier, allowing everything to flow freely through us.

CHAPTER 8

Cycle Scopes—Your Cycle at a Glance

To support your cycle charting process, on the first day of your bleed—or if you no longer bleed, on the day of the Dark Moon—work with and refer to, these fun and informative cycle scopes to help you start to recognize what's ACTUALLY going on with your body and your cycle. Back in 2015, I created the #sharemycycle hashtag, where I shared my feels and my experience each day. At the time, I got lots of comments from people ick-ed out by it, but now, thankfully, that hashtag has been shared thousands of times, and social media is full of people talking about their bloody period and sharing their monthly menstrual experiences.

What I share here are super-short, sweet, and sometimes slightly tongue-in-cheek daily cycle scopes for a 28-ish day cycle. This is based on charting my own cycle, and the cyclic energies that clients have shared with me. Please know that this may absolutely NOT be how it is for you.

I also know many, many people don't have 28-day cycles. I suggest letting these 28-day at-a-glance cycle "scopes" simply be a guide. Keep your own cycle journal so that you create your own cycle scopes that are specific to you, your body, and your own menstrual experience and wisdom.

Day 1

It's quite possible that on the first day of your bleed, your entire womb is falling out of your vagina, and you'll feel a combination of sore, sacred, and like you're feeling it ALL. Estrogen starts out at ground zero on day 1, so don't be surprised if from today through to day 3, you feel achy and tired—immerse yourself in self-care, radical rest, and preservation of the self-love kind.

Day 2

Rising estrogen and testosterone mean that you may feel your first signs of energy and good vibes starting to return. Hurrah. Don't be fooled, though, while you may wake up feeling like you could do ALL. THE. THINGS, your body will most likely lose steam by mid-morning. Instead of blowing today's teeny bit of energy doing stuff that can be put off, take it sl-o-o-o-o-w. If you've got any big questions or decisions to make, bleed on it. Make a den, get comfy, place your hand on your womb, meditate, and heart riff. Let your womb be your oracle.

Day 3

As your estrogen and testosterone continue to rise, around day 3 you're able to start refocusing and wanting to make sense of the world, but don't be in too much of a rush to "get back out there." Your flow may still be heavy, and you may not be able to fully articulate what you actually want to say.

Day 4

I always find that day 4 marks the start of "fun" me. Energy levels are rising, a sense of optimism and confidence returns. I'm able to turn my attention to the outside world again. Your blood may start to change color and texture around day 4 too. If you experience a heavy flow, it may be showing signs of getting lighter.

Day 5

If you experience a "normal" flow (whatever one of *those* is), your flow should start to feel lighter, as should your mood. Today's the day to get creative–think up solutions to problems, make art, tackle the emails, start to explore any ideas and downloads you've received during the last few days.

Day 6

Today you think AND feel like Wonder Woman. That creativity you were experiencing yesterday? Forget it. Today you're all about the logic. Your estrogen and testosterone levels mean you have clarity and focus, and by now, there's a chance you've also stopped bleeding too.

Day 7

The I-can-do-it-all vibes are HIGH, and you may notice a shift of energy as you move into the pre-ovulation phase. Tiny follicles on your ovaries are starting to form in preparation to release an egg, and while you're amped up and ready for anything, there's the potential for tiny dramas to turn into full-on anxiety-inducing stress-outs. You can blame estrogen for that.

Day 8

While you may be feeling practical, logical, and like a fully-functioning grown-up today, follicles are continuing to grow in prep for the egg release, and the egg development is causing your estrogen levels to rise. That rising estrogen? It's why, despite the adulting, you're tempted to book a last-minute flight to LA, get a tattoo, and stay out dancing until 2 a.m..

Day 9

I like day 9 a LOT. It's my "I'm in-control" day. Everything feels do-able here, as if I might actually have got my shit together. Ha! It's no surprise then that these hormones have us feeling like we've figured out adulting right at the same time that extra estrogen is thickening the lining of the uterus, getting it ready just in case an egg gets fertilized and needs a comfy place to grow for the next nine months. Our bodies are SMART.

Day 10

Ha! I laugh in the face of adulthood as estrogen is now sky-high, *and* testosterone is rising toward its peak and has me wanting to take ALL the risks and say "Yes" to EVERYTHING.

Day 11

From now through until day 13, hormonally, you're on fire. THESE are the days to work late, to get shit done, make decisions, and DO. ALL. THE. THINGS. If you're looking to conceive, these are also the days to ramp up the amount of sex you're having too. Wink.

Day 12

Sound the sirens, this girl is on fiyaaaahhhhhhhhhh! If you thought yesterday was good, today just gets better. Energy levels and your ability to articulate are high, you're sharp, funny, and you're good to be around. This is the day I schedule any social events–it's a very small window of opportunity when *this* particularly introverted extrovert wants to "people."

Day 13

Today you're a total powerhouse: testosterone and estrogen are both now at their peak, you're fertile, and if it's what your body does, it will release an egg. You'll be feeling the potency. If you're not making out or making a baby, use this potent Creatrix power to create something else–a book, a project, or a piece of art.

Day 14

So testosterone and estrogen have peaked, and for some, day 14 is ovulation–but this isn't true for everyone. Check your vaginal fluid to be sure. If there's a lot of it, and if it's egg white-y in look and consistency, then there's a good chance you're ovulating.

If you don't want to make a baby when your vag is producing the slippery stuff, you should deffo be using a barrier method, because THIS is the good stuff. When you're ovulating, you're magnetic, and your aura stretches to 26ft (that's an exaggeration, but y'know, it's big—use it).

Day 15

You'll still feel the good vibes from the last few days, but be prepared, as those hormone levels start to drop, so does your energy levels. The good news? Those energy levels don't drop right away, so if there's work to do, a project you need concentration, focus, and brainpower for, use today's energetics to get it done. That egg? It's now journeying down the fallopian tube looking for some sperm action.

Day 16

Heads up, this is your last day to access the ability to concentrate, focus, and do all of that logical-brain thinking because, from here on in, you're going to start to wish you hadn't said "Yes" to "doing" quite so many things in the days that follow. You'll feel less interested in socializing and more interested in having long baths filled with crystals, oils, and rose petals (wait, that might just be me?). Progesterone has entered and is encouraging you to become more self-nurturing and to choose comfy-ness over skydiving. Why? Because if there was sperm and egg action, it too will need nurturing and comfy-ness, NOT skydiving.

Day 17

So today may feel like one big reality check. You see everything exactly as it is, and that's not always easy to do. Your uterus lining is thickening in case there's potential baby action, estrogen is dropping, while progesterone is high. Cue tears. ALL. THE TEARS. Or rage. Or both.

Day 18

Testosterone and estrogen appear to have changed their minds and start to rise again, but this time without all of the feel-good vibes they brought with them earlier in the cycle. Boo. Why? Because progesterone is basically a buzz-kill. Its job is to make sure that if an egg was fertilized in the last few days, your womb stays safe.

Day 19

Despite testosterone and estrogen rising, progesterone is still in charge. That's all.

Day 20

OK, so today, right through until you bleed, trying to "make sense" of things starts to get a bit harder as you now move into emo, "all the feels" territory. If you're progesterone-sensitive, it may activate tears and confusion, and trying to do anything "practical" starts to get harder from here on in. I know, right? *Shakes fist at progesterone.*

Day 21

Between today and day 25, I'm not my most favorite person to be around. I try not to schedule meetings or anything that involves people if I can possibly help it, for these few days. In good news, the progesterone rising does make these few days good for clearing out, editing, revising–it's the only day in my cycle you'll catch me with a hoover. Actual fact.

Day 22

Ohhh, today, progesterone peaks. This means that you may need at least one (preferably 84 in my case) nanna naps to overcome hormone-fuelled fatigue. I, for one, will not judge. Mainly because I'll have been under my blanket napping too.

Day 23

ALL the hormones are now on the decrease. Of *course* they are. If you've any kind of big decision to make, put off making it until after you bleed if you can. It's not that you can't trust yourself here, but this second half of the cycle is new territory for many of us, so until you're practiced at making emotional decisions with discernment, I recommend waiting until you move back into day 6.

Day 24

Progesterone levels are still high in comparison to the rest of your hormones, which are plummeting about now. What does this mean? Well, for me, usually tears, rage, and chocolate. A lot of chocolate. Forever Cacao, my FAVOURITE Cacao supplier (see REDsources at the back of the book), are very well utilized between days 21 and 25.

Day 25

This has me showing up like Kali Ma incarnate. Rage-full and angry and without patience or tolerance. This is a tricky day for a lot of us as all the hormones are withdrawing at this point, BUT if channeled correctly, or if in my case, I'm kept away from people, it can be a really creative time.

Day 26

Personally, I always feel a bit better after day 25, but lots of my clients feel those day-25 vibes waaay into day 26 too. I feel like if you KNOW the emotions that show up here (for example, rage is a big one for me), and you allow them, they can then move through a lot more easily. But if we're holding onto them, pushing them down, trying to ignore them, or worried we might explode, they can create pain. It's why our creativity is heightened here so that we can make art/magic from the feels.

Day 27

Ohhh, thanks to those plummeting energy levels, everything may feel like it's in s l o o o o o w motion. You may be wanting to fight the many, many injustices in the world, but ultimately your energy levels make that a little difficult, so then you get frustrated, not just at your inability to be in action, but also at the state of the world, and so it goes. I often wake up wanting to eat ALL THE FOOD before 7 a.m. on day 27, so if you can make that food good and high in protein, not just family-size bags of chips—you'll thank me, I promise.

Day 28

If you can, stay away from social media, as your comparison and not-good-enough buttons can get REALLY pushed here, and the sharpness of your truth-telling tongue combined with the lack of brainpower to articulate totally effectively makes social media a not-so-great space to be. If you're not pregnant, there's a good chance that your period is on its way.

Tracking and working with your cycle isn't a prescription for how to be and act on each and every day. (I mean, it *can* be, but would you really want it to be?) Instead, see it as a way to explore and get curious about your body. Learn what it's up to and understand how, based on your cyclic self-knowledge, you can let it provide you with insight and wisdom as to why you feel like you do and how you "could" show up.

What happens now?

First, let me start with a massive high five and deep, deep bow. It takes ovaries to dare to get to know yourself better: YOU RULE.

Now the work can *really* begin.

Think of yourself as a member of the *Time Team* except instead of digging for ancient bones, you're digging for clues regarding you and your experience as a woman.

If I'd been taught about this cyclic wisdom—the phases of my cycle and how to chart them—from menarche, my first bleed, I would now have a full-to-the-brim, go-to guidebook to my moods, feelings, and emotions. I would know myself. I would have had fewer arguments with my parents, a better understanding of my friendship dynamics with other women, and, quite frankly, I'd have had a *much* better sex life in my 20s. So, don't waste a single minute: get to know *your* flow, and let the unlocking of your superpowers and your cyclic wisdom begin.

What affects your feelings and emotions, and, most importantly, when are they triggered? What cycle phase are you in when they are? Where's the Moon at?

Look to see if day 21, for example, evokes the same thoughts and feelings for you each month, and if so, what can you do to ease it or make it better?

I have a gorgeous friend who highlights her bleed days in her diary, puts an "out of office" on her email saying, "Please do not disturb, I'm bleeding," and literally takes four days off.

Each month.

Now, I know that's not possible for many of us—our cycle may not be so predictable, we may have work and family commitments—but imagine a day like that or even an hour.

Day 25 is an absolute non-negotiable day off in my world. It's not a bleed day, but it's a day that I've identified, through charting my cycle, as a day I need to NOT be around people. I won't coach clients, do interviews, meet with friends, go out, go online, or even answer the door because I know that I have a tendency to kick off, get mad, scream and shout a lot, and instead of trying to push it down, and repress it, I allow my Wise, Wild Woman a day to express herself.

I have cave time: I lay out my sheepskin rug, light candles, take my journal, a good book, a flask of tea—although if the Viking is home, he'll leave me tea and banana bread outside the door—and I allow myself to be fully in it. Sometimes it comes and I have to hit a pillow, but more often than not, because I've created a retreat space, because I've slowed down and dropped all my responsibilities, the anger doesn't arise. Instead, I just appreciate a day to read and chill out solo. Also, just so you know, the world doesn't end if you take a day out of your schedule. I've actually found that by taking day 25 off, issues and potential drama-rama that would have previously unfolded because of my actions are now non-existent.

Carving time out for yourself on the days that you need it the most, is *the* most incredible act of self-care. And, just so you know, self-care is NOT a luxury; it's a necessity. Make THAT your mantra as you explore your cycle.

If your cycle is more or less regular, check to see if there's a particular time in the month where you feel more powerful. This is good to know because if these feelings show up in the same place each month, you can put them to good use.

Get the idea?

The good news is that you don't have to have it all figured out right away, and personally, I spent over fifteen years of my life with my head in a self-help book or an online program of some description. The practice of getting to know my cycle has cut through all the bullshit, and I'm finally able to start making sense of why I've been feeling the way I've been feeling. I've been a tamed woman, and, through this practice, I now declare myself untamed.

As women, we're not short of sticks to beat ourselves with. There are the big generic societal ones, the ones given to us by partners/family members/work colleagues, and then there's the biggest and most painful sticks of all—the ones that we give ourselves and hit ourselves with repeatedly, on a daily basis. I want *Code Red* and the act of charting your cycle to show you that you ARE enough. Right now, you're trying to operate in a world that's simply not been set up for you—it's linear, clear-cut, and consistent, so every time you try to operate within it, you feel instantly bitch slapped with the "not good enough" and "I'm a failure" labels. When you know your flow, and when you access your superpowers accordingly, you're led directly to the magic that exists right inside of you, just waiting for you to discover it. You connect to a life that's more fluid and ever-changing—it doesn't go in straight lines or follow a set path. Your red brick path, Dorothy, has bends and turns and spirals, it ebbs and flows, and as you connect with your body and listen to the truth that emanates from inside you, you care much less for the bombardment of images and information that are sent on a moment-by-moment basis to distract and disconnect you. Instead, you're able to use your cycle to ground down into Mumma Earth and make firm roots wherever you are. Things are less able to shake you because you trust yourself and what you know.

Basically, you are bloody brilliant.

CHAPTER 9

Let's Hear It for the Boys

So where do men fit into all this?

The difficulty for men is that, until now, we ourselves haven't been fully aware of the power of our cycle. We haven't had the vocabulary or the knowledge to explain what's going on for us each month. We have known *something* was going on, especially in the second half of our cycle. If you've been particularly resistant to the power of your cycle, then there's a good chance you've experienced a lot of tamed and censored Wise, Wild Woman angst in the form of anger, anxiety, and stress. But without knowing what was going on ourselves, how could we explain it to the men in our lives? Yet, if you live with or are in a relationship with a dude, and you're anything like me, there's probably been an expectation for him to somehow "get it", and then total annoyance ensues when he inevitably says or does the wrong thing.

For the most part, men are walking a tightrope when it comes to women and their cycles.

The solution?

Know your cycle. Then share that knowledge with him.

SHE STORY

Sharing cycle charting with himself, by Tamara Protassow Adams

I've always been aware of cycles and felt connected to nature. In my childhood and teens, I wandered through our garden and the local bush solo, feeling the pulse of nature's rhythms, and observing the tiniest changes in leaves, plants, and the weather.

It had occurred to me that my nature was linked to my experience of the world, that my inner workings were as intricate as those of the natural world I'd so closely observed growing up. Still, the distractions of adult life, as a wife, mom of two, and business owner, just took precedence. Add to that a partner who saw "that time of the month" as an inconvenience at best—cue critical, weeping woman—and the bane of his existence at worst—cue critical, weeping woman—and I had a recipe for suppressing my true self, ignoring what could be the truest teacher I'd known.

A little background…

Himself, my husband, while being a believer in all sorts of spiritual and shamanic kookery, also had a big problem with my cycle. He just couldn't see why I had times when I was more negative or keen to sort through things that weren't working. Times that made me super-aware of the bad stuff that just needed to get gone from my life—often including him and how we related. As for the times I was more emotional, well, they were just too much. Riding waves of emotion have never been his favorite pastime, and the fact that this happened with (gasp!) monotonous regularity in his life with a wife just bewildered him.

When I started charting my cycle with attention, Himself was a bit bemused, but he also admitted to being fascinated, and more than a little at sea. His belief system includes an intense spirituality, but one that is a little more removed from everyday than mine. I believe that spirit and everyday are entwined, and contact with spirit is just a thought away. Charting my cycle and being aware of the practical, spiritual, mental, and emotional significance of the changes I was going through on an everyday level was a new concept for him. He really thought discounting the premenstrual phase as "hysterical" or "off your tree" was acceptable, no matter how many times I'd argued, wept, and accused him of not particularly caring for me because of this. It drove me bonkers!

Enter charting and the "Explore Your Lady Landscape" course with Lisa. I only ever showed Himself the first video in the series, where the four stages of the cycle are explained, and the value of deeply going into each one fully laid out. The premenstrual clarity and anti–bullshit–ometer were finally given a value and a place—and it changed the way Himself thought about my cycle enormously.

He could see the value of having a week where super–clarity was a go, and the rose–tinted spectacles (though awfully convenient and trippily nice) were thrown out. My challenge was to focus on things and situations other than him, but the big change was that it became normal. Normal in his mind, and therefore more normal for me, because I no longer needed to defend my own self—a full quarter of my experience—from his insistence that it was somehow wrong and defective.

Of course, I was also making some dietary changes which had been a long time coming, and they helped to stabilize my mood a bit. But by far the biggest, most significant shift for Himself was to an expectation that this was normal, and even to working with it. As an example, we were designing an extension to our home at this point. Himself took advantage of the fall part of my cycle to run changes past me and find out exactly what was "not–quite–right." He'd then spend a few weeks tweaking and redrawing the plans, then come back to me again the next time I was premenstrual to run them past the anti–bullshit–ometer again! It was so effective and a great way to use that cycle phase.

His attitude shift has needed the occasional reminder from me that being a woman is a biological experience that I have no choice in and no control over—in the essential, basic, this–is–my–body–and–I–can't–help–it sense. But the seeds of understanding have been

```
sown, and acceptance is more of a feature of our day-
to-day now than it ever has been.

The odd side effect is that now I can feel like I'm
having the PMS trip from hell, and when I tentatively
apologize for what I feel has been a hard few days,
he'll often say that he's not noticed terribly much
"clear-seeing, bullshit-busting" behavior. Acceptance
of it really has changed us both.
```

Some of my favorite cycle findings, which ultimately have been a deepening knowledge about myself, have been through my relationship with my husband, the Viking.

There's a delicious intimacy that comes from understanding yourself and your sexual/romantic wants and desires (and sharing this knowledge with your partner too). You may discover that during menstruation you need to pull away, or after menstruation, you need slower, more attentive loving, while at ovulation, you crave more passion and energy and want and need sex NOW. You may find that different parts of your body are more sensitive at different times in your cycle. Maybe you love your nipples being tweaked in your late pre-ovulation, but in your premenstrual phase? You shout, in no uncertain terms, "Step away from the nipples, I repeat, step away from the nipples!" Knowing this about yourself is so liberating, but what's more important is to communicate with your partner about it, get them involved in the charting process, and, most importantly, encourage them to explore, experiment, and get to know you and your body and your sexual preferences at each phase of your cycle. It really does make for an awesome relationship in and out, under and on top. Wink.

Now, it hasn't always been plain sailing here. Sometimes the Viking and I would argue, and there were times in my cycle when we were just out and out miscommunicating. I would get so frustrated. Basically, I was pissed that he didn't understand. I was about to bleed, my body was heavy, and I was grumpy, so no, I didn't want sex/to go out/play naked twister, OK? Then that would turn into an argument, which mainly involved me shouting and him sulking.

So, in one of these aforementioned "moments," he demanded I make him a guide to my moods. I think it was a sarcastic quip, but you should never joke with an art-girl about to get her period, so I drew out all the moods I found myself in throughout my cycle, cut out a li'l arrow to indicate the mood I'm in on that day, laminated it, and placed it beside his bed.

Now each morning before he gets up, I set the dial to my mood for the day.

Just so he knows. (It also comes with a caveat that it can change at any given moment, obvz.)

It was just a bit of fun, but, actually, it has been rather practical too, because our arguments have been significantly less since it's been a part of our relationship.

Due to the nature of my work, I talk, teach, and write about periods, vaginas, sex, wombs, ovaries, and fertility a LOT, so the Viking has no choice but to wholeheartedly get to know my cycle. When he first sat in on some of my workshops, he said, "You need to do a boy version of this. Men need to know this information so they can fully honor and respect the women in their lives. Also, knowing what day your wife or partner is going to go all Kali Ma on your ass really helps in deciding whether to stay in or go work the late shift and leave cookie jar outside the bedroom door."

Totally relevant man logic at its best.

But there would be no point in me writing a guide for men. I'm not one, so I suggested he write one instead, and he did, and it's good.

So, leave the book open at this page and urge any boy or man you know to read it—don't worry, they can thank us later.

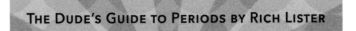

THE DUDE'S GUIDE TO PERIODS BY RICH LISTER

Right, dudes, I've asked for a guide like this, and there's isn't one. A guide that explains how women work and what I can do as a dude to make her happy/have an easy life. Consider this the holy grail of woman. (Because, really, what you're about to find out is the holy grail.)

Your job is to read this and not stick it behind the fridge or otherwise ignore it, but to read it and then talk to your love about it. Then share it with your friends, teach it to your sons, and don't be as afraid to talk about it with your daughters. You don't have to talk about the sex bit if that feels uncomfy, but do you know what? They'll need to know that too, and it's best you tell them than they get their information from porn, surely?

Women are governed by cycles–not bikes but the seasons, the Moon, and menstrual cycles. Each month, for roughly a 28-day period–from their first bleed through to menopause–their hormone levels rise and fall, and this changes how they act and show up in the world. You might notice that the woman in your life may have very subtle changes; others may find that the women in their world ride a rollercoaster of hormones and emotions during their monthly cycle. Don't worry, there's no science bit, just think of the four seasons–spring, summer, fall, winter. Women travel through and experience each season once a month, and as they travel through each season, their mood, their needs, their likes and dislikes change. So, each month you are spending time with four different women.

Your job is to simply know which season she's in, what moods and temperaments are associated with that season, and what you can do to meet her needs and honor her (and to score major partner points in the process).

Unfortunately, there isn't a one-size-fits-all failsafe formula because each woman is unique (thank the goddess for that), but work with your partner as she charts her cycle and learns what works for her. Take notes if you have to because if you understand this, your life will be better.

She bleeds, right?

Your life gets complicated when she does, right?

You like sex, right?

You will WANT to know how this works.

So, my wife talks about periods a lot. Part of her work is to teach women to chart their period and how they feel throughout it. Because she's a geek, my wife has been doing this for years and has folders and folders of charts, which basically are a manual to how she works. For three months, we charted together. We made notes on her moods, her needs, her sexual preferences, when she wanted to be left alone, and a whole lot of other really important stuff.

What was interesting was that I already knew parts of her cycle. I knew roughly what day she was at too. I knew that when she was crying on the stairs or shouting at me for not taking out the garbage, she would be hours away from starting to bleed. Or that for a couple of days before her bleed, it was probably a good idea for me to work long shifts and to come home with presents. Preferably chocolate.

All of this is important.

Not for just keeping me out of trouble, but for making our lives together better.

What wisdom would I impart?

Things that have been important discoveries through this process that have made my life a LOT better:

When to get laid.

When to fuck up.

When to date.

When to cuddle.

When to go out with my friends.

When to stay home and cook dinner.

When to tidy the house.

When to have an argument.

When to win an argument.

The things that will make your life better may vary; in the same way every woman varies. Your job is simply to learn how your woman flows. Here follows a breakdown of each phase and what to look out for:

Pre-ovulation/spring–days 7 to 13

Spring is when she has just finished bleeding, the pads/tampons/cup/sponges have gone away, and she's starting to come out of her cave.

Yay, right?

Yup, she will want cuddles and to be stroked, and this is a great time to woo her or take her out on dates. Don't try and have sex right away–you may have blue balls, but no. If you want sex, you're out of luck, making out is the order of the day just after her bleed.

To get laid: include lots of foreplay.

Lots.

Flowers, cuddles, passionate kisses. You'll have to work for it. She'll be soft and tender, and there's a good chance she'll want you to cuddle after, so don't fall asleep straight away.

Going out with your friends is good, but make sure you text her what's going on. She's just coming out of the delicate part of her cycle, so may need a little reassurance, but don't fret too much as she'll be booking dates with her friends too.

There's a good chance you'll be able to watch the rugby, but be prepared to have her cuddled up next to you with her laptop, looking at social media and asking, "Who are the team with the blue shorts?"

Ovulation/summer–days 14 to 20

In the transition from spring to summer, her hormones are going to say, "Yes!" She'll be hot, intoxicating, and awesome to be around. She'll be fun, and, yes, she'll be feeling horny.

Now is the time to book adventures, to go away for a long weekend, order room service, and mess up the sheets.

The good news is that during spring and summer, you could paint the house the wrong color, accidentally wash her whites with your red socks, or buy a new car and not get in too much trouble. Basically, you can cock up and not get your head bitten off quite so much at this time of the month as you might at any other.

Summer is the "do it" part of her cycle.

Want babies? Now will be the time to try. Not wanting babies? Use contraception.

Premenstruation/fall—days 21 to 28

Sex will need a lot more foreplay because her erogenous zones will be a little less sensitive. Don't take this to mean that it's a green light on the nipple clamps. Ask first, yeah?

She'll want to watch her favorite movie on repeat and eat her favorite food. Make this happen. She'll want to nest, so make sure you have tidied up your sports kit from the hallway and that the dog is walked. This will save a lot of arguments because, honestly, you're not likely to win many at this time of her cycle, so my advice is to do everything you can to make sure they don't happen.

Toward the end of fall, about day 25, I personally go to work and stay late. I say "Yes" to every 14-hour shift because I'm able to do no right. It's not her fault; this part of the cycle is calling her to slow down and rest, but if she doesn't, it can make her a little difficult to reason with.

Your job here is to find out what she needs, so talk to her. Don't lead with: "You're so moody, you must be premenstrual"–that won't end well–but ask if there's anything she needs and let her know you're there if she needs you. But, honestly, this is a good time to book in some time with your friends, and just ask her to call you if she needs anything.

Also, beware, she'll want to go over everything. Multiple times. When are the utility bills due? Have you rung your mom? Is the cat too fat? All the things.

Menstruation/winter—days 1 to 6

The Bodyform advert lied.

You know that blue-balled feeling you get when all you want to do is orgasm? Imagine that feeling for FOUR days. That's what she's been feeling in the days leading up to her bleed, so when the blood comes, it's a big release.

Now, she will bleed for days.

She won't die, phew, but she'll be tired. She'll also need more protein in her diet, and probably won't want to do too much, so keep plans on the down-low.

If you go out with your friends, bring home a present, or a good story or gossip. Just don't come home empty-handed.

Sex? Probably best to wait until day 3 or later as the blood flow will have lessened, and she'll be soft, tender, and want loving. If she does want sex during her period, don't be icked-out by it—it's awesome.

Offer to cook dinner and give her space when she needs it through her bleed days. You'll get a lot of kudos.

This is just a guide. What you experience with the woman in your life may be different. Your job, being in a relationship with her, is to learn how to work with her. Talk about it, talk about your needs throughout this too. What do you need? If, like me, you need lists, ask for them, but by knowing what each other want and need and, most importantly, talking about it, you'll find a whole new love and respect for each other.

Now go forth and chart cycles together. Argue less. Make out more. Love each other hard.

NOTE

For those in same-sex couples, or if you're in a couple and you both menstruate, or if you're a woman sharing your living space with other women, don't assume that cohabiting will instantly mean your cycles sync, or that you'll experience each phase of the cycle in exactly the same way. You might, but you also might not. My advice? Chart your own cycle, get familiar with how it is for you, then share your experiences. Get familiar with each other's patterns and cyclic tendencies in each phase so that you can make living with each other way more harmonious, respectful, and love-filled.

My Final Wish for You

Before I began this path, I had a very fixed idea of what it was to be a woman, and for most of my youth, I rebelled against it: cutting my hair short, wearing boy's clothes, and preferring to play soccer and hockey rather than join the netball team.

If we don't think we fit a particular ideal—and I most definitely wasn't the just-stepped-out-of-a-salon, picture-perfect, slim-with-a-flawless-complexion girl who flicked her hair, which I held up as the image of so-called perfection when I was at school—it lowers our self-esteem and how we value ourselves.

When I began to recognize and, most importantly, accept that a woman can be many, many different things, it allowed me to shake off limiting stereotypes and find the areas in life where *I* rocked and ruled, and then show up in the world on *my* terms. When you drop the labels and accept there are many faces of the feminine, and you can pick any or none of them, you can then begin to love, nurture, and really accept yourself.

Heal your menstrual cycle

By accepting and expressing the different facets of who you are as a woman, you start the process of living your cycles consciously. You begin to recognize who you are and what you need at different times in your menstrual cycle and in the cycle of your life.

If you've previously felt a lot of shame and blame attached to menstruation, or when you were unpacking your menarche story you found some big dark shadows, take them out, get curious about them, familiarize yourself with them, wrestle them to the ground, and have it out with them if you have to. If anything you discover about yourself feels too big or too overwhelming, book a SHE session with me, or take it to a trained therapist. Don't feel you have to figure it out on your own.

By understanding your cyclic nature, and experiencing your rising and falling energy levels throughout your cycle, you'll start to appreciate your body and give it exactly what it needs.

Knowing there are certain times of the month when your body needs to rest and renew (and get the benefit of reduced menstrual symptoms) means you can begin to accept your body and not fight it. This self-knowledge and body acceptance is priceless.

You are SHE-rarr—not to be confused with the '80s cartoon heroine *She-Ra*, although both are equally as good—you're a divine force, and you've the power to know, work with, and maximize each phase of your cycle in every area of your life. Let's rewrite the story of what it means to be a woman in the world, and let's start that story with you.

Forget mastery, this is *mistress-ry*.

Big, big love,

Lisa x

Deep Bows and High Fives

The biggest love + gratitude to everyone who has shared their wisdom and expertise with me in this book—you are golden—thank you for supporting us all with your medicine.

Each and every woman who has ever shared her stories, her fears, her greatest wishes with me in workshops, in classes, in immersions—deep, deep bows of respect.

And to YOU— for holding this book + daring to enter in—you are bloody brilliant.

REDsources

Free download

Cycle tracking charts and playlist: Download yours by visiting www.thesassyshe.com/coderedthebook and using the password: yesibleed

My bookshelf

(Because everyone loves looking at someone else's bookshelves, right? That CANNOT just be me.)

Amberston, Celu. *Blessings of the Blood: A Book of Menstrual Lore and Rituals for Women*

Angier, Natalie. *Woman: An Intimate Geography*

Beak, Sera. *Red Hot and Holy: A Heretic's Love Story*

Block, Francesca Lia. *Guarding the Moon: A Mother's First Year*

Diamant, Anita. *The Red Tent*

Dinsmore-Tuli, Uma. *Yoni Shakti: A Woman's Guide to Power and Freedom Through Yoga and Tantra*

Eisler, Riane T. *The Chalice and the Blade: Our History, Our Future*

Estés, Clarissa Pinkola. *Women Who Run with the Wolves: Myths and Stories of the Wild Woman Archetype*

Gates, Janice. *Yogini: The Power of Women in Yoga*

Grahn, Judy. *Blood, Bread, and Roses: How Menstruation Created the World*

Grigg-Spall, Holly. *Sweetening the Pill: or How We Got Hooked on Hormonal Birth Control*

Hill, Maisie. *Period Power: Harness Your Hormones and Get Your Cycle Working for You*

Iyengar, Geeta. *Yoga: A Gem for Women*

Kent, Tami Lynn. *Wild Feminine: Finding Power, Spirit & Joy in the Female Body*

Knight, Chris. *Blood Relations: Menstruation and the Origins of Culture*

Lark, Susan M. *Menstrual Cramps: Self-Help Book*

Moran, Caitlin. *How to Be a Woman*

Monk Kidd, Sue. *The Secret Life of Bees*

Musicio, Inga. *Cunt: A Declaration of Independence*

Harding, M. Esther. *Women's Mysteries: Ancient and Modern*

Northrup, Christiane Dr. *Women's Bodies, Women's Wisdom*

—— *Mother-Daughter Wisdom: Creating a Legacy of Physical and Emotional Health*

Owen, Lara. *Honouring Menstruation: A Time of Self-renewal*

Pope, Alexandra and Bennett, Jane. *The Pill: Are You Sure It's For You?*

Pope, Alexandra. *The Wild Genie: The Healing Power of Menstruation*

Rako, M.D., Susan. *No More Periods?: The Risks of Menstrual Suppression and Other Cutting-Edge Issues about Hormones and Women's Health*

Raymond, Janice G. *Women As Wombs: Reproductive Technologies and the Battle Over Women's Freedom*

Richardson, Ani. *Nutrition for a Happy, Healthy Period: Positively Managing Symptoms (of PMS) with Nutrition and Lifestyle Changes*

Richardson, Ani. *Love or Diet: Nurture Yourself Release the Need to Be Comforted by Food*

Shuttle, Penelope and Redgrove, Peter. *The Wise Wound*

Stepanich, Kisma K. *Sister Moon Lodge: The Power & Mystery of Menstruation*

Roth, Gabrielle with Loudon, John. *Maps to Ecstasy: The Healing Power of Movement*

Tiwari, Maya. *Women's Power to Heal: Through Inner Medicine*

Vitti, Alisa. *Womancode: Perfect Your Cycle, Amplify Your Fertility, Supercharge Your Sex Drive and Become a Power Source*

Walker, Barbara G. *The Woman's Encyclopedia of Myths and Secrets*

Watterson, Meggan. *Reveal: A Sacred Manual for Getting Spiritually Naked*

Weed, Susun S. *Healing Wise: The Wise Woman Herbal*

Welch, Claudia Dr. *Balance Your Hormones, Balance Your Life: Achieving Optimal Health and Wellness through Ayurveda, Chinese Medicine, and Western Science*

Wolfe, David. *Naked Chocolate: The Astonishing Truth about the World's Greatest Food*

Online resources

Thesassyshe.com: Supporting women to heal and explore their relationship with their body and their power as they navigate these "interesting" times.

Mpowder.store: Information and products for perimenopause, menopause, and post-menopause.

kennyethanjones.com: The first trans man to front a period campaign and have important, open conversations about menstruation and body politics.

Floralunity.com: The amazing Salvatore shares THE best flower essences to support sensitive souls and has an amazing *Code Red* essence kit for each phase of the menstrual cycle.

Em Tivey: A visionary herbalist and medicine maker, email her at emmativey@gmail.com for consults and medicine-making.

forevercacao.co.uk: Pablo and Tad are my go-to chocolate dealers! These two super-gorgeous humans work with care and love to support the communities from which their cacao is ethically sourced. They provide ceremonial-grade cacao and also make drinking chocolate and cacao bars infused with rose, which I have been known to consume in one sitting on day 1 of my cycle.

honouryourflow.co.uk: An independent online supplier of handmade cloth pads in a range of gloriously vibrant colors. I also really love gladrags.com and lunapads.com.

rootandflower.co.uk: I cannot recommend ALL of these products enough. Raw, organic, 100 per cent natural remedies and skincare all made in small batches using traditional artisan techniques with total love. I share here because Jen has created a *Comfort: Woman* blend to harmonize the emotional and physical imbalances felt during menstruation, and it is ACTUAL magic.

ABOUT THE AUTHOR

Lisa Lister, author of *Love Your Lady Landscape* and *Witch*, is a well-woman therapist, yoga teacher, and menstrual maven.

She is the Creatrix of IN-YOUR-BODY-MENT—a combination of yoga, sound, breathwork, and somatic movement that works in collaboration with the wisdom of the cycles: body, seasons, nature, and the cosmos.

Lisa offers practical, psychological, and spiritual tools, guidance, and support to women who are exploring, navigating, and healing their relationship with their body and their cyclic nature; and asking the question: what does it mean to be a woman in these "interesting" times?

 @sassylisalister

www.thesassyshe.com

Listen. Learn. Transform.

Listen to the audio version of this book for FREE!

Today, life is more hectic than ever—so you deserve on-demand and on-the-go solutions that inspire growth, center your mind, and support your well-being.

Introducing the *Hay House Unlimited Audio* mobile app. Now you can listen to this book (and countless others)—without having to restructure your day.

With your membership, you can:

- Enjoy over 30,000 hours of audio from your favorite authors.
- Explore audiobooks, meditations, Hay House Radio episodes, podcasts, and more.
- Listen anytime and anywhere with offline listening.
- Access exclusive audios you won't find anywhere else.

Try FREE for 7 days!

Visit **hayhouse.com/unlimited** to start your free trial and get one step closer to living your best life.

HAY HOUSE
Look within

Join the conversation about latest products,
events, exclusive offers and more.

f   Hay House

🐦   @HayHouseUK

📷   @hayhouseuk

🖤   healyourlife.com

We'd love to hear from you!